When Computing Got Personal

A history of the desktop computer

Matt Nicholson

matt
publishing

BRISTOL ENGLAND

First published in Great Britain by
Matt Publishing
7 Unity Street
Bristol, BS1 5HH
http://blog.mattmags.com

ISBN-13: 978-0-9927774-1-8
Edition 1, February 2014

Cover design by Colleen Lanchester-Raynie
Cover image supplied by istockphoto.com

To Hazel, Jemma and Luke

Contents

Introduction

In December 1983, I joined VNU Business Publications in London to become editor of *What Micro* magazine, the first consumer buyers' guide in the UK to tackle the new market for what had become known as microcomputers. I had studied Computer Science at university, but that had been six years earlier on room-sized computers fed by paper tape and punched cards. Microcomputers were new to me, so that Christmas I took home a guide to the MS-DOS operating system and a WordStar manual in the hope that they would help me understand what was going on, and how to use the ACT Sirius 1 that sat on my desk in place of a typewriter. The dim green cursor on its otherwise blank screen was blinking in anticipation of me entering something intelligent, and I didn't want to disappoint.

However, what I found in those guides baffled me. Like so many I had assumed that computers, being logical machines, would function in a logical fashion. Instead I found a hodgepodge of seemingly unrelated concepts and illogical key sequences. It was only later that I realised my mistake. The machines themselves may be logical – even maddeningly literal at times – but the way they work is a function of their design, and they were designed by human beings who made decisions that were not always rational, that may have reflected compromises made years earlier and no longer relevant, or were attempts to standardise or improve on what went before. In short, if I was to really understand not only how computers work, but why they work the way they do, I would have to understand their history.

A full history of computing would reach back into the 19th century, when Charles Babbage conceived his Analytical Engine and Ada Lovelace started dreaming up programs for it. However, as the title suggests, this history is concerned with the computer as a personal device, and with the people that made it possible for an ordinary individual to own and use one. As such, its starting point is the development of the integrated circuit in the 1960s, and the appearance of the first microprocessors in the early 1970s.

Winston Churchill is alleged to have said, "History is written by the victors", and indeed much of this book is concerned with those who could be viewed as winners: people like Bill Gates and Steve Jobs, and companies like Intel, Microsoft and Apple. However, just as important are the stories of those that didn't make it. As I hope this book will demonstrate, the difference between winning and losing is not simply down to making the right or the wrong decision. Quite often, those who we now perceive as having lost made the only decision they could, given their circumstances and what they knew at the time. It's also easy to underestimate the role of chance, and the way that one company's decision can seal the fate of another.

But it's not all business and politics. It is also a showcase of human ingenuity, driving an industry that in just four decades turned room-sized computers into devices many magnitudes more powerful and yet small enough to fit in your pocket. Here you will find some of the most intelligent, and some of the most eccentric, people you could hope to meet, including wide-eyed hippies, subversive students, computer nerds, entrepreneurs, hackers, crackers and financial backers. Some will lose and some will win, but all play a part in an adventure that has transformed, and continues to transform, our world.

There are thousands of stories relating to the development of the personal computer that could be told. I have chosen to include those that tell us something important about the way the industry developed, and can be backed up by reliable sources. Inevitably much has been left out, and some may argue that the book is incomplete or prejudiced as a result. In particular, it could be said that I have concentrated too much on America, at the expense of what was happening in the rest of the world. There is coverage of events that took place in both Britain and Japan, but my response would be that most of the significant developments did take place in America, partly because of the huge investments made by the US government into research and development, and partly because of the sheer size of its economy.

Readers should be aware that, although the early chapters are essentially sequential, later chapters tell stories that unfold in parallel. I have therefore included a timeline towards the back of the book which lists the more significant events in chronological order. I have tried to avoid unnecessary technical detail, and to include explanations within the text where appropriate. However, readers can also refer to the section 'How computers work' for a brief description of underlying concepts.

1.   Minis and mainframes

In 1972, Alan Kay, a scientist who worked at Xerox's Palo Alto Research Center, presented a paper at a conference in Boston with the title 'A Personal Computer for Children of All Ages'. In it he proposed a "personal, portable information manipulator" that could be used by both children and adults, which he called a Dynabook. The paper was unusual in that it opened with a short story which has two nine-year-olds named Jimmy and Beth sitting on the grass in a park near their home with Dynabooks on their laps, playing *Spacewar* across some sort of wireless network. Later Jimmy connects to the Library, where he becomes "heir to the thought and knowledge of ages past", while Beth's dad is sitting on a plane, "perusing pertinent background facts" on his Dynabook in preparation for the coming business meeting. Earlier, his eye had been attracted by "a lurid poster on one of the airport's StoryVends". After he downloaded the racy text, "he was chagrined to be reminded by the StoryVend that he had neglected to pay for the copy" (Kay, 1972).

In these days of notebook PCs, tablets and smartphones, it's easy to forget how far-fetched such ideas seemed in those days, just 40 years ago. Computers were in common use, but in a fashion very different to that of today. Most large organisations possessed a single mainframe so big that it occupied a large air-conditioned room, and so complex that it could only be operated by trained technicians. Ordinary 'end users' were granted access through dumb terminals that were housed separately. Alternatively some university departments, research centres and the like had their own mini-computers. Certainly no individual, except the exceedingly eccentric and exceedingly rich, would consider actually owning one.

Even in science fiction such a possibility was rarely raised, although the Tricorder used by crew members in the original *Star Trek* series of the late 1960s does exhibit some of the computing power of the modern smartphone. A more typical vision is to be found in the film *2001: A Space Odyssey*, released in 1968 and featuring the HAL 9000, a single computer that controlled, and indeed was an integral part of, the spaceship Discovery One on its voyage to the planet Jupiter. Although the film has the HAL 9000 becoming operational in 1992, it exhibited a level of image recognition and overall intelligence that is well beyond anything we can achieve today.

The basic design of almost all computers was, and indeed still is, much the same as that outlined in 1945 by John von Neumann in his *First Draft of a Report on the EDVAC*. In this report, Neumann divides the computer into four component parts: an arithmetical unit that performs the actual operations; a memory unit that stores both the sequence of instructions that the computer is to perform (the 'program') and the data on which it is to act; a control unit that orchestrates the processing of the instructions; and an input/output unit that connects the computer to the outside world, allowing us to enter information and view results.

Early computers were set up for batch processing. Anyone wanting to use the computer would first have to enter the program and the data into a keypunch device, which looked a little like a typewriter but generated either a punched tape or a stack of cards. The keypunch was not connected to the computer itself, so the user would hand the roll of paper tape or the card stack to a computer operator, who would feed it into the machine at their convenience. The computer would then process the instructions and print out the result ready for your collection at a later time (or date, if the computer was particularly busy).

By the 1960s, most computers were using transistors rather than thermionic valves, making them much faster. As a result, timesharing became possible, allowing many users to access the computer simultaneously from remote teleprinters or terminals with the illusion of real-time interaction. This was achieved by having the computer divert its attention from one program to another in a 'round robin' fashion, perhaps 100 times a second. Although the computer was only ever doing one thing at a time, timesharing created the illusion that it was doing many things at once. The trade-off was in speed: if ten users were connected then they would each experience a computer running at a tenth of its full capability – or indeed slower as the hardware would have to spend some time saving the data generated by one user before loading the data needed by the next. As a consequence, most computer scientists preferred to work at night when there were fewer users, giving them a bigger slice of the computer's time. Computer bureaux (companies that rented out access to a timesharing computer) tended to charge less at night in an effort to spread the load.

Dominating the computer industry since its inception was the American company International Business Machines (IBM), which had launched its

System/360 mainframe in 1964 – the year that it celebrated its 50th birthday – following an investment programme that *Fortune* magazine dubbed "IBM's $5 billion gamble". By this time, IBM was well on the way to becoming the largest company in the world, with nearly 150,000 employees and pre-tax profits of over $400 million (equivalent to nearly $3 billion today). It held over 70 per cent of the market, the remainder being distributed between its competitors Burroughs, Control Data, General Electric, Honeywell, NCR, RCA and UNIVAC, a group commonly known as 'the Seven Dwarfs'.

The System/360 was built using a precursor of the integrated circuit known as Solid Logic Technology, developed by IBM for this purpose. The machine was available in many different configurations, all very impressive for the time but with less memory and a slower processor than you would find in a modern mobile phone.

Mainframe computers were primarily used for data and transaction processing. An early success for IBM was SABRE, short for Semi-Automated Business Research Environment, which in 1964 took over all of the booking requirements for American Airlines. SABRE used 12,000 miles of cable to connect 1,100 reservation desks in 60 cities across America to two IBM 7090 mainframes (a precursor of the System/360) which were located in New York State. The software became the basis of the Customer Information Control System (CICS) which is still used by many financial institutions today. Most companies did not buy a mainframe computer, but instead rented it as part of a package that included software and maintenance. The original press release for the System/360 quotes monthly rents ranging from $2,700 to $115,000 for a "large multisystem configuration".

Some would say that IBM's dominance in the marketplace was inevitable – even desirable. Any company taking delivery of a mainframe entered into a lifelong relationship that could only become stronger. IBM technicians continued to adapt the system to better suit the company's needs, while IBM salesmen ensured that the customer was up to date with the latest options. Such a relationship was very hard for a competitor to break, even if they could offer a lower price, and indeed through the latter part of the 1960s, the proportion of IBM customers who actually defected to another manufacturer was a fraction of a per cent (DeLamartar, 1986, p. 72). Many of its competitors resorted to building equipment that was compatible with an IBM machine, and so could be used alongside it. However IBM had only to change the interface of its machine and the competitor would have to go back to the drawing board while IBM filled the gap with one of its own products.

As a result, the company frequently stood accused of illegal and unfair business practices. Indeed IBM's first president Thomas Watson Sr. had been a defendant in the first case ever to be brought under America's anti-trust laws back in 1912, receiving a jail sentence as a result of his tactics while working for the National Cash Register Company (NCR). The Roosevelt administration had filed another suit in 1932 that accused IBM of entering into a conspiracy with its chief competitor to monopolise the market for blank punched cards. Then, in 1952, the government launched yet another anti-trust suit forcing the company to change its business practices. Finally, the Johnson administration filed a suit in 1969 charging IBM of illegal activities on numerous counts, a case that dragged on for 13 years until the Reagan administration finally withdrew, claiming the case was "without merit".

Although IBM continued to dominate the market for large mainframes, a new breed of smaller 'minicomputers' started to appear in the early 1960s, made possible as first transistors and then integrated circuits came into widespread use. Legend has it that the term was coined by John Leng, head of the UK office of Digital Equipment Corporation (DEC), when he opened a sales report with: "Here is the latest minicomputer activity in the land of miniskirts as I drive around in my Mini Minor" (Jones, 2001). Particularly popular was the PDP-8 which was launched by DEC in 1965 from its headquarters in the small town of Maynard, Massachusetts. The acronym stands for Programmed Data Processor as the company's investors felt that the word 'computer' implied that it was big and expensive. Unlike the mainframes of the time, the PDP-8 was the size of a refrigerator and did not need to be housed in a separate room. Prices started at $16,200 for a basic configuration with a memory capable of storing 4,096 binary 'words' of 12 bits each. This was for purchase rather than rental, making it much more affordable and highly popular, selling some 50,000 units. It was followed in 1970 by the PDP-11 which was even cheaper at $10,800 (equivalent to a little over $50,000 today) and proved even more popular.

Meanwhile, computers were attracting the attention of a new generation. Universities had been offering degree courses in Computer Science since the early 1960s, and it was not unusual to see a minicomputer tucked away in the corner of a laboratory. Computer programming had become easier thanks to BASIC

(Beginner's All-purpose Symbolic Instruction Code), a computer language developed in 1964 by John Kemeny and Thomas Kurtz at Dartmouth College, New Hampshire with the intention of helping all students make better use of the computers that the college owned, regardless of the subject they were studying.

Ludwig Braun, who had long been convinced of the importance of computing to education, was a big fan of BASIC which he felt to be far better suited to the needs of ordinary people than the FORTRAN language favoured by the industry itself. In 1967, he managed to persuade the National Science Foundation (NSF) to fund an exploration into the subject at the Polytechnic Institute of Brooklyn. For the Huntington Project (so called because it initially worked with schools in the Long Island suburb of Huntington), Braun brought together some 80 high school teachers, introduced them to BASIC, and then opened up a discussion into how programs written in BASIC might help in the teaching of their various disciplines, which ranged from biology and physics to history and social studies.

The result was a series of simulation programs, many of which proved to be forerunners of games to come. One of the first was *POLUT*, which simulated the effect of pollution on a body of water. Students could choose whether this was a pond or a slow or fast-moving stream, the nature of the pollutant, the rate of injection and whether the water was warm or cold. The teletype would then print a graph showing how the levels of waste and oxygen in the water changed as each day passed, and alerting students when conditions deteriorated to the extent that fish could no longer survive. At this point, Braun reported, the students would react "with great sadness, and they went back to the drawing board determined to do something to keep the fish alive" (Johnstone, 2003, p. 57).

The Project proved so successful that in 1970 the NSF funded a second phase which became known as Huntington II. This resulted in the publication by DEC of teaching manuals for 17 such simulations, each accompanied by a paper tape copy of the program code, which were to be used by some 600 teachers working with more than 25,000 pupils (Visich & Braun, 1974, p. 3). Other simulations included *MARKET*, which simulated two companies competing to sell a product, and *CHARGE*, which simulated Robert Millikan's 1909 oil drop experiment to measure the electric charge of an electron.

Young enthusiasts were also getting together informally after school. One particularly well-known group was the RESISTORS, which stood for 'Radically Emphatic Students Interested in Science, Technology and Other Research

Subjects'. The RESISTORS was formed in 1966 by a group of teenagers (both boys and girls) from Hopewell Valley Central High School in New Jersey who spent much of their time tinkering with out-dated computers that had been donated to the group, including a Burroughs 205 valve-based computer from the early 1950s which weighed nine tons. Later, as the RESISTORS became better known, DEC donated a PDP-8. Typical activities included a program called the Conceptual Typewriter which displayed shapes on an early visual display in response to key presses. It was created for *SOFTWARE*, an exhibition held in 1970 at the Jewish Museum in New York and subtitled 'Information technology: its new meaning for art' (RESISTORS, 2009).

A similar exhibition had been held in 1968 at the Institute of Contemporary Arts (ICA) in London, which the author attended as a teenager. Exhibits at *Cybernetic Serendipity: the computer and the arts* included computer generated poems, pictures and films. The composer Herbert Brün had used an IBM mainframe to generate a short musical piece entitled *Stalks and Trees and Drops and Clouds* (because that's what the notation looked like), while Jeanne Beaman and Paul Le Vasseur of the University of Pittsburgh had programmed a similar machine to choreograph a dance.

Students had been writing and playing computer games since the beginning of the decade. One of the earliest was *Spacewar,* which was written in 1961 by Steve 'Slug' Russell to run on the PDP-1 minicomputer that DEC had donated to the Research Laboratory of Electronics at Massachusetts Institute of Technology (MIT). *Spacewar* involved two players manoeuvring spaceships around the gravity well of a central star while firing streams of torpedoes at each other. The action was depicted on the PDP-1's circular monochrome screen, and the game played using two makeshift controllers. Although very simplistic it proved immensely addictive, and Russell's fellow students were soon busy improving and enhancing the game. Copies of the paper tape containing the *Spacewar* program were widely circulated, to the extent that engineers at DEC were using it as a final diagnostics check to confirm that a PDP-1 was working properly before delivery (Levy, 2010, p. 56).

The students of many universities have long-standing traditions of conducting pranks, but few are as long-standing or elaborate as those carried out by students at MIT, which have long been referred to as 'hacks'. Halloween of 1962, for example, saw the Great Pumpkin hack, which involved decorating the university's famous

Great Dome with strategically placed pieces of black cloth and orange gel so that it looked like a huge pumpkin complete with eyes, nose and a crooked grin.

It is therefore not surprising that students with access to the campus computers should be thinking about their potential for mischief. In particular, one Stewart Nelson had started as a freshman at MIT in 1963 and had quickly fallen under the spell of the PDP-1. Another student, Peter Samson, had written a program called the *Harmony Compiler* that played music through a small loudspeaker. Inspired by this, Nelson placed the loudspeaker close to a telephone and wrote a program that could control the phone system, generating tone sequences to search for external lines and establish long-distance connections free of charge. The story eventually found its way into the student newspaper and is often cited as the first use in print of the term 'hackers' with reference to computers:

> Many telephone services have been curtailed because of so-called hackers, according to Professor Carlton Tucker, administrator of the Institute phone system. Stating "It means the students who are doing this are depriving the rest of you of privileges you otherwise might have," Prof. Tucker noted that two or three students are expelled each year for abuses on the phone system. The hackers have accomplished such things as tying up all the tie-lines between Harvard and MIT, or making long-distance calls by charging them to a local radar installation. One method involved connecting the PDP-1 computer to the phone system to search the lines until a dial tone, indicating an outside line, was found. (Lichstein, 1963)

Computers were beginning to be seen not just as tools for business but as instruments of creativity, entertainment and even rebellion.

2.   Cheap as chips

What Kay described back in 1972 foresaw wireless networks, the World Wide Web and even online shopping, and has only recently become a reality with the likes of the modern notebook PC or tablet. However, as Kay emphasised in his paper, the full benefits of the Dynabook could only be realised if it was cheap enough for ordinary people to own. Only when millions of them were in use would technologies like wireless networking and the World Wide Web make sense. The Abstract of his paper opens with the following:

> This note speculates about the emergence of personal, portable information manipulators and their effects when used by both children and adults. Although it should be read as science fiction, current trends in miniaturization and price reduction almost guarantee that many of the notions discussed will actually happen in the near future. (Kay, 1972)

At the time he was writing, integrated circuits had been around for a while. As he realised, they would prove crucial to the realisation of his vision.

Before the integrated circuit or 'silicon chip', electronic circuits were constructed by wiring discrete components together, usually by hand. As the complexity of the circuits increased, more and more connections had to be made between more and more components, which made assembly time consuming and prone to error (a problem known as 'the tyranny of numbers'). This was acceptable while only a few computers were being built and each could be sold for large sums of money, but not if computers were to become cheaper and more common. The integrated circuit allowed complete electronic circuits, with transistors, diodes, resistors and so forth, to be built into the surface of a single silicon chip, and while the initial cost of research, development and testing was high, the cost of duplication was relatively low, so the more chips were made, the cheaper each one became. Furthermore, the greater the number of components that could be crammed onto the surface of each chip, the fewer the number of external connections that had to be wired by hand, and the lower the cost of final assembly.

The integrated circuit (IC) is the result of two inventions, one by Jack Kilby of Texas Instruments in Dallas and the other by Robert Noyce, a founder of Fairchild

Semiconductor which was based in Palo Alto, one of the larger cities in the Santa Clara Valley in southern California (now known as Silicon Valley). What Kilby had realised was that resistors and capacitors could be constructed from the same material as transistors and diodes, and in 1958 he demonstrated a device that put a complete electronic circuit, comprising a pair of transistors and a number of other components, onto a slice of germanium measuring just 1cm by 0.2cm (Kilby, 1959).

Noyce chose silicon rather than germanium as the semiconductor. His contribution was to use silicon oxide as an insulator and a metal coating process to create connections, so making it easier to manufacture smaller devices containing a greater number of components (Noyce, 1959). As he wrote in his notebook:

> It would be desirable to make multiple devices on a single piece of silicon, in order to be able to make interconnections between devices as part of the manufacturing process. (Slater, 1987, p. 158)

Both Kilby and Noyce patented their inventions, and the two companies were in legal dispute for some years, until in 1966 they agreed to cross-license their respective technologies. By then other companies had entered the market, including Motorola and Signetics (established in 1961 by a group of engineers from Fairchild and also based in Silicon Valley).

At first, ICs were primarily used in high-budget government projects where space and weight were at a real premium, such as the Apollo Guidance Computer, designed at MIT for the moon landings, and the guidance system used by the Minuteman II intercontinental nuclear missile (ICBM) programme. These large-scale projects gave American companies the resources they needed to ramp up production and so bring unit costs down, which gave them an advantage over foreign competitors. Semiconductors were being manufactured in Japan by companies like Nippon Electric Company (NEC), Hitachi, Toshiba and Matsushita; and in Britain by Plessey, General Electric Company (GEC), Ferranti and Mullard. But by 1964, Texas Instruments was itself responsible for around a third of global semiconductor production (Anchordoguy, 1989, p. 28), while by 1968, Texas Instruments (Bedford) Ltd had become Britain's largest semiconductor manufacturer (Morris, 1990, p. 114).

By the mid-1960s the technology existed to cram 50 to 100 transistors onto a single chip, and products such as TI's popular SN7400 series, which implemented various logic functions in standard 14-pin 'dual-in-line' packages, were available at competitive prices and finding their way into both mini and mainframe computers. However Pat Haggerty, president of Texas Instruments, was of the opinion that the market needed a 'demonstration product': something that would fire the imagination and boost demand. Finding himself on a flight one day in 1965 seated next to Jack Kilby, he suggested to Kilby that integrated circuits would allow them to develop a calculator as powerful as the desktop models available at the time, but small enough to fit in your pocket. Kilby was able to demonstrate a prototype to a group of colleagues a year later, and filed for patent in 1967. The prototype used seven ICs which between them contained nearly 5,000 transistors.

It was not Haggerty's intention at that time for TI to start manufacturing calculators, so the company entered into a partnership with the Japanese company Canon, supplying the chips for a consumer version of Kilby's prototype called the Canon Pocketronic. Launched in 1970, the Pocketronic went on sale in the United States in 1971 at a price of $395 (equivalent to around $2,000 today). Other companies were entering into similar arrangements, so in 1970, Sanyo launched its ICC-0082 'Mini Calculator' using chips made under a licensing agreement with American company General Instrument. The Sharp QT-8B 'Cordless Micro Compet' was launched in Japan in the same year, using ICs supplied by North American Rockwell Microelectronics.

By the end of the 1960s, the economies of scale had taken hold to the extent that digital ICs were being advertised in hobbyist magazines such as *Popular Electronics* in the US and *Practical Electronics* in the UK at prices that even a teenager could afford. A 1968 catalogue from the popular mail order supplier Radio Shack advertises a Dual JK Flip-Flop chip for $2.49 with the line, "Construct your own binary computers, adding machines etc.". By 1972, it was offering a pack of 20 assorted digital ICs for $2.99 (about $14 today). Hobbyists and electronics enthusiasts who had been building radio receivers and audio amplifiers out of first valves and then transistors were beginning to experiment with digital components.

Furthermore, the April 1965 edition of the trade magazine *Electronics* had featured an article by Gordon Moore, a co-founder of Fairchild with Noyce, predicting that integrated circuits containing 1,000 components would be cost effective by 1970. He continued:

> The complexity for minimum component costs has increased at a rate of roughly a factor of two per year ... Certainly over the short term this rate can be expected to continue, if not to increase. Over the longer term, the rate of increase

is a bit more uncertain, although there is no reason to believe it will not remain nearly constant for at least 10 years. That means by 1975, the number of components per integrated circuit for minimum cost will be 65,000. (Moore D. G., 1965)

This remarkably prescient statement became known as Moore's Law. Ten years later, Moore altered his projection to suggest that from 1980 onwards, the number of components it would be economical to put on a single chip would double every two years (Moore D. G., 1975); a statement which remains substantially true today, with the latest microprocessors packing over a billion transistors onto a chip of silicon just 1.5 centimetres across.

Early digital ICs packed a handful of simple logic circuits into a single chip. However as the number of components that could be crammed in increased, new options were opening up. The first memory chips appeared, with capacities increasing from 256 to 2,048 bits of data by 1970. Such chips came in two forms: Random Access Memory (RAM), where the data stored can be changed with ease but is lost when the power source is removed; and Read Only Memory (ROM) where the data is physically encoded into the chip and cannot easily be changed. Manufacturers were beginning to contemplate the possibility of building processing units, or even a complete computer, from just a handful of silicon chips.

Fairchild was finding it hard to compete with Texas Instruments in this market, and most of its original founders had left to start new ventures, many within Silicon Valley. In 1968, Noyce and Moore left to form a new company down the road in Santa Clara which they named Intel, an abbreviation of 'Integrated electronics'. Rather than compete directly with their old employer, they decided, initially at least, to focus on manufacturing memory chips. The following year they were approached by Nippon Calculating Machine (NCM) to develop a set of chips for its Busicom range of desktop calculators.

NCM's approach was similar to that of other calculator companies, in that it called for 12 discrete purpose-built ICs. However semiconductor manufacturers were beginning to realise that a general purpose design, which could be programmed to suit the needs of each customer, would make their lives much easier. Such a system would combine a central processing unit (CPU) with various control and memory components. It could then be customised by physically writing the necessary program instructions into the system's ROM, a process known as mask programming which involves 'burning' the program code into the ROM chip during manufacture. The underlying architecture would be that of a von Neumann machine, while the programs became known as 'firmware'.

When Ted Hoff, manager of Intel's Application Research group, saw the Busicom specification, he realised that it could be achieved in a much more flexible fashion using just four chips, including a 4-bit CPU which could be programmed to execute specific tasks (the section 'How computers work' gives a more detailed explanation of these terms). Busicom agreed to go ahead with Hoff's design and the contract was signed in February 1970, but work didn't really get under way until April when Frederico Faggin took over as project leader, having left Fairchild to join Intel. Faggin brought with him the silicon gate technology that he had developed at Fairchild which would greatly facilitate the construction of higher-density chips, and give Intel a crucial advantage over Texas Instruments.

Meanwhile another company, the Texas-based Computer Terminal Corporation (CTC), had approached Intel with an even more challenging proposition. CTC had come up with the idea of a computer terminal that could be programmed to work with any mainframe computer, simply by loading a specific terminal emulation program through its built-in tape drive. The Datapoint 2200, as the terminal was to be called, boasted a full QWERTY keyboard, a monochrome screen that could display 12 lines of 80 characters each, an 8-bit CPU (necessary for handling ASCII characters), two tape cassette drives for storing data and programs, and 2KB of RAM that could be expanded to 16KB. In all but name, the Datapoint 2200 was a desktop computer, and indeed once in production it was widely used as a standalone computer system. However CTC was promoting it as a terminal to a mainframe computer.

CTC was already a regular customer of Intel, so the meeting was concerned with the chips that would be needed for the Datapoint 2200, and in particular how to construct its 8-bit processor. CTC had already designed the CPU's instruction set, and knew that it could be implemented using around 100 standard logic chips. However, they also wanted the 2200 to be small – ideally the same size as IBM's popular Selectric typewriter – so they wanted to use as few chips as possible (Poor, 2004, p. 32). It is not clear who first came up with the idea, but the upshot of the meeting was a decision to implement the 8-bit processor as a single chip (Faggin, Feeney, Gelbach, Hoff, Mazor, & Smith, 2006, p. 2). Intel started work on the 1201, as the chip was initially named, at around the same time as work began in earnest on the 4-bit processor for Busicom.

CTC was also a customer of Texas Instruments, and in early 1970 the two companies agreed that TI should develop a single-chip processor to the same specification. The TMX1795, which was TI's internal name for the project, would provide CTC with a second source should Intel have problems delivering the 1201. However by the middle of the year, CTC had decided to go ahead with its original plan to implement the CPU using discrete logic chips. The price of standard logic ICs was coming down, making such an approach more economical, and CTC reasoned that it could always replace them with a single chip processor at a later date.

Intel put the Datapoint project on the back-burner and Faggin focused his attention back on the 4-bit chip for the Busicom calculator. Texas Instruments continued to work on the TMX1795, and eventually presented CTC with a working version in July 1971. However it did not function as well as the processor that CTC had built out of discrete logic chips, and so was not used. Nevertheless Gary Boone of TI filed a patent for a 'Computer Systems CPU' the following month, based on his work developing the TMX1795 (Boone, 1973). The patent was granted in 1973.

Meanwhile, Seiko had heard about the 1201 and expressed an interest in using it in a forthcoming scientific calculator, prompting Intel to revive the project. At the time, Intel was in two minds about the microprocessor as a product line. It had already built up a successful business selling memory chips to computer companies, and could count Burroughs, Control Data, IBM and UNIVAC amongst its customers. Many within Intel, particularly the sales team, were wary of announcing a range of microprocessors that would effectively put the company in competition with its main customers. However others, such as Faggin, disagreed; as Ted Hoff put it, "Eventually, somebody else is gonna open up this business and we'll be left behind, even though we're the first ones to have products" (Faggin, Feeney, Gelbach, Hoff, Mazor, & Smith, 2006, p. 11).

In fact, computer companies did not have a problem with microprocessors, as they saw them being used as programmable controllers for mainframe peripherals such as tape drives or printers, rather than powering computers in their own right. Indeed most within the industry saw these devices as microcontrollers rather than microprocessors – chips that could be mask-programmed to perform specific tasks such as controlling household appliances, scientific devices or industrial machines.

And, in fact, such a view has been borne out by history. Today the average household has far more microprocessors embedded in its cars and TVs than it does in its desktops and laptops. The personal computer, where users can change the program that the processor is running with relative ease, represents only a tiny portion of the market.

By early 1971, Intel was able to deliver working versions of the 4-bit processor to NCM in Japan, who used it for the Busicom 141-PF calculator. Launched in April 1971, this was the first calculator to employ a single-chip CPU. However NCM was finding the calculator market very competitive, and the agreement it had drawn up with Intel did include provisions for the $60,000 development fee to be returned in the event that Intel wanted to use the design elsewhere (Intel, NCM, 1970). Faggin and Hoff persuaded Noyce that Intel should exercise this option, with the result that the company made the four Busicom chips available as the MCS-4 Micro Computer Set, taking out an advertisement in the November 1971 issue of *Electronic News* under the banner: "Announcing a New Era in Integrated Electronics: a Microprogrammable Computer on a Chip!"

NCM didn't do so well and ceased trading three years later. Nevertheless the Busicom name continues, having been bought by the sole importer of the Busicom range in the UK, Broughtons of Bristol, which still sells a range of office machines, including desktop and pocket calculators.

At the centre of the MCS-4 was the 4004, a single chip 4-bit CPU containing 2,300 transistors and able to execute programs at a rate of 60,000 instructions a second. These instructions could be stored in the 4001, a separate ROM chip with a capacity of 256 bytes. The 4002 chip provided just 80 four-bit words of RAM, while the 4003 handled the interface with external devices such as keyboard and display.

Although primitive by today's standards, this was a very flexible package. At its most basic, Intel could supply just the 4004 microprocessor together with a single 4001, mask-programmed to the customer's specification. Alternatively the processor could be combined with multiple 4001, 4002 and 4003 chips to create a computer with up to 32,768 bits of ROM and 5,120 bits of RAM that could, for example, add 1,000 eight-digit numbers together in less than a second. The 4004 itself was initially priced at $200 (equivalent to around $1,000 today).

In the same month, Intel completed work on the 8-bit 1201 microprocessor. As CTC no longer wished to use it, Intel agreed to drop its $50,000 development fee in exchange for full rights to the design, including the instruction set originally drawn up by CTC. Intel renamed the chip the 8008, branding it as the obvious successor to the 4004, and launched it in April 1972 as the centrepiece of the MCS-8 product family.

The 8008 contained 3,500 transistors, ran at a clock speed of 500KHz (five times faster than the 4004) and could directly address up to 16KB of memory – considered at the time more than enough for any conceivable application. It was programmed to execute a set of 48 instructions, largely based on the original CTC specification, while its 8-bit data bus allowed it to deal with eight bits of data in one go, which made it eminently suitable for handling alphanumeric data and also easy to use with many of Intel's existing RAM and ROM chips.

To help people get to grips with its new microprocessor, Intel put together the SIM8-01, which it described as a "complete byte-oriented computing system". This placed the 8008 on a printed circuit board measuring 29 by 24 centimetres, together with up to 16KB of memory and the circuitry necessary to interface with devices such as a standard teletype unit with a keyboard and display, or a paper tape reader. By this time, Intel had also produced the first Erasable Programmable ROM (EPROM) in the 1702. Invented by Israeli engineer Dov Frohman, this 256-byte chip could have its contents erased by ultraviolet light and had a transparent window on top for this purpose. New content could then be written electrically onto the chip, and Intel provided a device for both writing and erasing 1702 content.

In the meantime the Datapoint 2200 was selling well, to the extent that CTC changed its name to Datapoint Corporation and by the early 1980s had become a Fortune 500 company. However, following a run-in with the US Securities and Exchange Commission (SEC) over dubious accounting practices, and a hostile takeover in 1985, the company eventually filed for bankruptcy in 2000. What remains of Datapoint is now based in Europe and sells call centre systems.

Even while working on the 8008, Faggin was thinking of ways in which the design could be improved. One of his problems was Intel's insistence on confining microprocessors to the 16 or 18-pin packages it was using for its memory chips, which helped keep the cost down. Moving to a 40-pin package would give Faggin far greater scope, allowing him to dedicate 8 pins to the data bus and a full 16 pins to the address bus, so obviating the need to multiplex data or address transfers and allowing direct access to 64KB of memory (see the section 'How computers work').

New technology also meant that the circuitry could be more tightly packed which made higher clock speeds possible, allowing it to process instructions at a faster rate.

Management was still to be convinced of this new market, so it was not until the summer of 1972, once the MCS-4 and the MCS-8 had proved popular, that Intel finally gave Faggin the go-ahead to implement his ideas. The result was the 8080 which finally went into production in March 1974, ushering in a new world. The 8080 contained 4,500 transistors and ran at a clock speed of 2MHz, while its ability to directly address up to 64KB of memory without multiplexing made it far more efficient than its predecessor. It was made available at a unit price of $360 (legend has it that the figure was inspired by IBM's System/360). As Faggin put it, "The 8080 really created the microprocessor market. The 4004 and 8008 suggested it, but the 8080 made it real" (Faggin F. , 1992).

Intel was obviously a major player right from the start, but the question of who actually invented the microprocessor remains open to question. The Intel 4004 was certainly the first microprocessor to go into mass production, and was officially announced in November 1971. However by then, Boone had already filed his patent for a 'Computer Systems CPU', while Gilbert Hyatt of Micro Computer Incorporated (MCI), a short-lived company whose investors included Intel's Noyce and Moore, had filed a patent entitled 'Single Chip Integrated Circuit Computer Architecture' back in December 1970, at around the time that Intel was working on the 4004 and the 8008 (Antonoff, 1991).

It was 20 years before Hyatt's patent was eventually granted, by which time the microprocessor had grown into a multi-million dollar industry. However in 1996, Boone brought a 'patent interference' case against Hyatt, with the result that Hyatt's claim was invalidated on the basis that key aspects of the description were lacking in his original 1970 filing. It was also agreed that Boone should be granted a Statutory Invention Registration (SIR) instead of a full patent, recognising him as the inventor but preventing him from collecting royalties.

Other contenders include the F14A Central Air Data Computer (CADC), designed by Ray Holt and Steve Geller for the US Navy's 'Tom Cat' fighter and manufactured by American Microsystems of Santa Clara in June 1970. This used six chips to implement CPU and memory, as opposed to the four used by Intel's MCS-4. However issues of national security prevented them from publishing details of their invention for nearly 30 years (Holt, 1998). Then there was the 8-bit

AL1 microprocessor developed by Lee Boysel, who left Fairchild shortly after Noyce and Moore to co-found Four-Phase Systems. The AL1 was manufactured by Four-Phase Systems and used in the company's System IV/70 display processor for mainframe computers, which was launched in 1970. However the AL1 was never sold as a product in its own right (Schaller, 2004, p. 314).

Although Intel lodged a number of patents during this period, it did not file one specifically covering the 4004 or the 8008. As Gordon Moore explains:

> ... frankly, we didn't think the microprocessor per se was that patentable. What we had done was take a computer architecture and make it all on one chip instead of on several chips. And that was kind of the direction that the integrated circuit technology was pushing in anyhow, always putting more and more of the system on a chip. (Interview with Gordon E. Moore, 1995)

As a business proposition, the microprocessor only makes sense as a mass-market product, allowing the huge costs involved in its initial design and development to be spread across thousands or even millions of units. Just what this meant was hard to grasp in the early 1970s: one story has a technical journalist asking how you would go about repairing such a chip, to which Hoff replied, "It's like a light bulb. When it's broken, you unplug it, throw it away and plug in another." Or to quote Stan Mazor, who worked on the 4004 and the 8008 with Faggin and Hoff: "The only thing that was significant about the microprocessor was that it was cheap! People now miss this point entirely" (Schaller, 2004, p. 301). What Intel had yet to discover was the full nature of that market, and just how big it would turn out to be.

3.　　The rise of the nerds

The early 1970s saw the younger generation basking in the aftermath of the Summer of Love. This was a generation that had become increasingly disillusioned with the materialism of the 1950s and 1960s. Underground newspapers were circulating revolutionary manifestos and protesting against the Vietnam War. Like London with its Kings Road and Carnaby Street, San Francisco had become a centre for the counter-culture. Alternative bookshops and purveyors of hippy paraphernalia had sprung up, particularly in the Haight Ashbury area. However the wider San Francisco Bay Area also encompassed Silicon Valley where companies like Intel, Hewlett-Packard and Xerox were taking their first tentative steps into a brave new world.

Corporate America, as symbolised by companies like IBM, ruled the world of mainframes and minicomputers, but these upstarts were in the business of rendering them redundant. The concept of a personal computer that individuals might own alongside a TV and a hi-fi was gaining credibility; and in the meantime, the timesharing facilities offered by existing machines opened up opportunities for ordinary people to experience computing for themselves. Members of the counter-culture were beginning to explore the social and political implications of this technology, and there was a growing feeling that, as Ted Nelson put it in *Computer Lib* (first published in 1974):

> You can and must understand computers NOW … Guardianship of the computer can no longer be left to a priesthood. (Nelson, 1987, p. 5)

Typical of such initiatives was Resource One, a non-profit collective operating out of a warehouse in San Francisco. Resource One had managed to persuade Transamerica Corporation to lend them a recently retired Xerox XDS-940 timesharing minicomputer, which they intended to use for various community-based projects. The most successful was Community Memory, instigated in 1972 by Lee Felsenstein and Efrem Lipkin, who were members of a group calling themselves Loving Grace Cybernetics (the name originates from a collection of poems by Richard Brautigan called 'All Watched Over by Machines of Loving

Grace' which was published in 1967) (Roszak, 2000).

Community Memory connected the XDS-940 to a teletype console sitting by the entrance of Leopold's Records in Berkeley. The connection was achieved using acoustic modems over a telephone line and managed just 10 characters a second. Anyone could write a message on the teletype, attach keywords and search for messages that used the same keyword, making it the first computerised bulletin board available to the general public. Despite its slow speed it soon contained thousands of messages offering items for sale, looking for jobs or simply commenting on the state of the world. It was an early precursor of the Internet café, and was soon joined by a second terminal at the public library in San Francisco's Mission district.

Then there was Bob Albrecht, a former computer engineer with Control Data. Like Braun of the Huntington Project, Albrecht had become convinced of the value of simulation games to education, and had spent time in the 1960s travelling to schools where he ran workshops for teachers. Albrecht moved to the Bay Area in the late 1960s and became involved with the Portola Institute in Menlo Park, which went on to publish the *Whole Earth Catalog* (an illustrated catalogue of products and services for the 'counterculture'). Albrecht and LeRoy Finkel, a teacher he had met on the road, continued to evangelise across California. As Albrecht recalled:

> We'd go to Hewlett-Packard and DEC to collect the computers they loaned to us, drive in my red VW bus to, let's say, UC San Diego, teach all weekend, pack up the computers, and drive home. (Johnstone, 2003, p. 67)

In October 1972 Albrecht, together with Finkel and a handful of others, launched the People's Computer Company with a newsletter that proclaimed on its front page:

> Computers are mostly used against people instead of for people; used to control people instead of to free them. Time to change all that – we need a PEOPLE'S COMPUTER COMPANY. (People's Computer Company)

Issues of the newsletter appeared every few months and contained a mix of news, tutorials and program listings, interspersed with the occasional music or film review and plenty of hand-drawn cartoons in the style of the *Whole Earth Catalog*. Central to the content were listings of games and simulations written in BASIC, which the newsletter proclaimed to be "the people's language". Many of these were based on programs from Braun's Huntington II project. Others included an early adventure game called *Hunt the Wumpus*; an intergalactic trading simulation called *Star Trader*; and *Kingdom*, a game of resource management. The listings for these games were later collected into the book *What to Do After You Hit Return*, which was published in 1975 with the assistance of Hewlett-Packard.

Another product of the People's Computer Company came courtesy of the PDP-8 that Albrecht had persuaded DEC to give him in return for writing *My Computer Likes Me When I Speak in BASIC* (Levy, 2010, p. 168). This was the People's Computer Center, a shop-front facility in Menlo Park where anyone could drop in and make use of the teletype terminals that were connected to the minicomputer, or remotely to another computer at Hewlett-Packard. Enthusiasts were standing by to show people what to do and demonstrate the programs that were available.

The first microcomputers (as computers built around microprocessors were becoming known) were based around the Intel 8008 processor and its accompanying SIM8-01 board. Intel itself introduced the Intellec 8 in 1973 as a tool to help its major customers develop software. Although not sold to the general public – at $2,400 it cost as much as a new car – the Intellec range did set the style for many of the microcomputers that were to follow. It did not come with a keyboard or screen, but instead boasted rows of switches and lights that were used to directly enter and display the binary content of the system's memory. However it did have an 'expansion bus' into which users could slot as many as 18 additional boards. Each board could contain up to 16KB of memory, or the electronics necessary to drive peripheral devices such as a tape reader or a teletype or one of the new 8-inch magnetic floppy disk drives that had recently been introduced by IBM.

Programming such a device required a great deal of patience. Most users did not have access to a floppy disk drive, which meant writing your program on a teleprinter and reading it into the microcomputer using a paper tape reader. Entering a 4KB program, consisting of perhaps 5,000 lines of machine code, could take several hours – and perhaps the best part of a day if the machinery was playing up. There were occasions when errors could be corrected by writing modified code directly into memory, but these were the exception. More likely was that the whole program would have to be entered all over again.

The Micral N, launched in February 1973 by the French company R2E (Réalisations Études Électroniques), is widely recognised as the first microcomputer to be sold to the general public. In addition to the 8008 processor and 2KB of memory, it boasted the Pluribus, an expansion bus that could take up to 14 additional cards. Originally built for the Institut National de la Recherche Agronomique (INRA) as a process controller, it was made available commercially at a price of 8,500 Francs – equivalent at the time to around $1,800, or $8,000 today. According to André Thi Truong, one of the company founders, R2E sold some 500 that year, and in May 1974 exhibited the Micral G, a successor which was based on the Intel 8080 processor, at the National Computer Conference in Chicago (Simpson, 1997).

Other microcomputers that used the 8008 processor included the SCELBI-8H from SCELBI Computer Consulting in Connecticut (the initials stood for SCientific, ELectronic and BIological). This was sold in kit form through a series of advertisements that appeared first in the March 1974 issue of amateur radio magazine *QST*, and later in *Radio-Electronics* magazine. The basic kit came with 1KB of memory at a price of $565, and a further 15KB of memory could be bought for $2,760. Unfortunately company founder Nat Wadsworth suffered two heart attacks during this period, at the age of just 30, which prevented the device from reaching its full potential. Wadsworth states that the company sold some 200 microcomputers over the next year or so, of which half were assembled and half sold in kit form.

However, while recovering in hospital, Wadsworth started writing a book called *Machine Language Programming for the 8008 and Similar Microcomputers*, which the company went on to publish. To his surprise it sold some 1,500 copies within a couple of months, which at $20 each made them more profitable than the computer. It was followed by titles such as *SCELBI 8080 Software Gourmet Guide and Cook Book* and *SCELBI's First Book of Computer Games for the 8008/8080*, which were similarly successful and led to the company abandoning hardware in favour of publishing (Gray, 1984) These 'cook books' contained complete programs listed in assembly language, which made them immensely popular amongst microcomputer owners looking for things to do with their new toys. The listings could take an hour or so to enter, and there was always the possibility of miss-typing or (worse) misprints, but until cassette and floppy disk drives became commonplace, such books were the only affordable way of distributing software to hobbyists.

Just a few months after the SCELBI advert, the front cover of the July 1974 issue of *Radio-Electronics* boasted 'Build the Mark-8, Your Personal Minicomputer'. Written by Jon Titus, the article described the construction of a computer that he had designed, based around a modified version of Intel's SIM8-01 board. Although never sold, either in kit form or assembled, readers could buy a 50-page construction guide for $5, which included board layout diagrams, or the full set of six ready-made printed circuit boards for $47.50. The article listed the other components required, including the 8008 processor, which at the time cost $120 from Intel. According to Titus, the magazine sold some 7,500 booklets and around 400 printed circuit boards (Johnson, 2009).

Not to be outdone by its rival, the editorial staff of *Popular Electronics* started looking for something even more spectacular to put on their front cover. This led to discussions with Ed Roberts, president of MITS (Micro Instrumentation and Telemetry Systems), a small company located just off Route 66 in a seedy part of Albuquerque, New Mexico. MITS had made its money selling electronic calculator kits, and indeed, its early success had been boosted by an article describing one of its kits that had been published in *Popular Electronics* a couple of years earlier. However ready-made calculators could now be bought for less than the price of his kits, which was causing his company serious problems. Roberts was looking for new ideas, and his interest had been piqued by the article on the Mark-8. He proposed to offer *Popular Electronics* readers a complete microcomputer kit for less than $400, including not only the instructions and the printed circuit boards, but all the necessary components as well. Furthermore, he wanted to use the new 8080 processor which Intel had only released a few months earlier. This was considerably more powerful that the 8008 used by the Mark-8, and would make the *Popular Electronics* kit even more attractive.

The problem was that the Intel 8080 cost over $300 when bought individually. Undeterred, Roberts persuaded Intel to sell him the 8080 in bulk for $75 apiece, on the basis that he accepted chips with cosmetic defects – damage to the housing or the chip surface that did not affect electrical operation. Roberts would have to sell 200 kits before he recouped his costs, which was a gamble, even given the magazine's readership of 450,000. After all, he was asking people to part with the equivalent of $1,700 in today's money for something which didn't actually do very much, even if they did manage to get it working.

Nevertheless, Roberts was able to ship a working prototype of the computer, which he had arbitrarily dubbed the PE-8, to the *Popular Electronics* offices that October. As a result, the cover of the January 1975 issue boasted 'World's First Minicomputer Kit to Rival Commercial Models'. Not keen on the rather bland name, the magazine's technical director Les Solomon asked his daughter for suggestions as to what they should name the device. She was watching *Star Trek* on television at the time and so replied, "Why don't you call it Altair? That's where the Enterprise is going in this episode." Accordingly it was branded the Altair 8800 (Solomon, 1984).

What you got for your $439, or $621 ready-assembled, was an Intel 8080 processor coupled to just 256 bytes of memory and a Display/Control board which allowed you to examine and modify the contents of any memory location using the 36 lights and 24 switches on the front panel. Roberts had also managed to secure a supply of 100-pin edge connectors which he used to implement an expansion bus into which you could plug expansion boards that made your computer more useful. MITS itself advertised memory boards at prices ranging from $176 for 1KB in kit form to $338 for a 4KB board ready assembled, together with a board that would allow you to connect a teleprinter, giving you a keyboard, a paper tape reader and a printer. A popular model was the Teletype ASR-33, but these cost around $1,000 and could only read 10 characters a second. Also mentioned was a 32-character alphanumeric display, an ASCII keyboard, an audio cassette tape interface and a floppy disk storage system, all of which were at the planning stage only.

Over the next few months following publication, MITS received orders for more than 2,000 units, which the company struggled to meet. As Les Solomon commented:

> The response to the article was totally unbelievable. What bothered me personally I guess was why would anybody in their right mind send $400 to an unknown company in Albuquerque, New Mexico, for a product that nobody really knew very much about. (Saxby, 1990, p. 159)

Unable to fulfil orders within the promised 30 days, MITS started offering refunds, but no one wanted their money back – they were all more interested in getting their computer, no matter how long they had to wait (Veit, 2002). The revolution had begun.

One person who bought the January issue of *Popular Electronics* was Paul Allen, who spotted it in a magazine rack as he walked across Harvard Square in

Cambridge, Massachusetts on his way to visit his old school friend, Bill Gates. Allen was a college dropout and computer geek, and the two of them had secured jobs working at Honeywell in Boston through the summer of 1974. With the start of the autumn term, Gates had gone back to his studies at Harvard, but spent most of his time playing poker and fooling around with the University's PDP-10 minicomputer. Gates was 19; Allen was 21.

When not playing *Spacewar,* Gates and Allen occupied themselves writing programs, and had become very familiar with the BASIC computer language. They immediately realised that a version of BASIC running on the Altair 8800 would make the device accessible to a much wider audience, and so phoned Roberts at MITS) claiming to have already written a BASIC interpreter that would work with the Intel 8080 processor. Roberts had also realised the importance of implementing BASIC on his machine, but was sceptical that a useful version could run in such a limited amount of memory. The two friends followed up with a letter offering to license their version of BASIC to Roberts in exchange for royalties on any copies that MITS sold. Such an arrangement did not involve any financial commitment on his part, so Roberts did not discourage them.

For Gates and Allen, the task was made more difficult by the fact that they did not have access to the machine itself, as the only working Altair 8800 was the one in the MITS office in Albuquerque nearly 2,000 miles away. However Allen had become very proficient with the MACRO-10 Assembler that ran on the PDP-10, and had previously spent time playing with the Intel 8008 processor. Allen's job was therefore to write an assembly program for the PDP-10 which would emulate the 8080 processor used by the Altair. All he had to work from was the *Popular Electronics* article and a technical manual for the 8080 which they found in a local electronics shop.

Meanwhile Gates had to write a BASIC interpreter in 8080 machine code that would fit into 4KB of memory, and still leave enough space to store the user's program and any variables that it used. In the end he managed to get a usable interpreter into 3.2KB, including floating point math routines contributed by fellow student Monte Davidoff, which left just enough room for the user to enter and run a BASIC program of around 50 lines of code. He later described it as "the coolest program I ever wrote" (Wallace & Erickson, 1992, p. 76).

Gates had told Roberts that they would have the interpreter ready in four weeks, but in the event it was eight weeks later that Allen flew down to Albuquerque with the finished code on paper tape. Both he and Gates were somewhat apprehensive as this would be the first time their program would be run on a genuine Altair 8800: up until now, their only means of testing it had been to run it on the Harvard PDP-10 against Allen's emulation program. To make matters worse, it was only during the flight that Allen realised they had yet to write the 'bootstrap' program that would actually tell the Altair how to load the interpreter into memory and execute it. Thankfully he managed to jot down the necessary machine code before the plane touched down.

The following morning he put the paper tape into the tape reader that was attached to the Altair in Roberts' office, and manually entered the bootstrap program using the toggle switches on the machine's front panel. There followed some anxious minutes until, to his immense relief, the attached teletype responded with the word 'READY'. Allen typed 'PRINT 2 + 2' on the keyboard, and the Altair responded by printing '4'.

Later that morning he managed to get the machine to run a listing of the Lunar Lander game which he found in a copy of *101 BASIC Computer Games*, a book published by DEC in 1973 to demonstrate the abilities of the PDP-10 (Kass). As Gates recalled:

> He and Roberts … were amazed that this thing worked. Paul was amazed that our part had worked, and Ed was amazed that his hardware worked, and here it was doing something even useful. And Paul called me up and it was very, very exciting. (Gates B. , 1993)

Shortly after the demonstration, Allen joined MITS as Software Director. Gates went back to his studies at Harvard but returned to Albuquerque for the summer break, when they continued to work on BASIC, starting development of an 8KB version that would offer more features. Gates and Allen had consolidated their partnership that April by creating the company Micro-Soft (the hyphen was dropped the following year), and in July the new company formalised its agreement with MITS through a ten-year contract drawn up with the help of Gates' father, an attorney with a law firm in his home town of Seattle, Washington.

Interest in these machines was growing, particularly amongst those who frequented places like Resource One or the People's Computer Center in Menlo Park. A few people had actually built microcomputers from component parts; many more were awaiting delivery of their kits from MITS with increasing frustration. Early that year, Fred Moore and Gordon French, both regulars at the People's Computer Center, decided to establish a club for those who shared such interests. The first meeting was held in French's garage on a rainy evening in March 1975, with just over 30 people from the Bay Area attending including Bob Albrecht, who was one of the founders of the People's Computer Center, and Lee Felsenstein of Community Memory fame. Albrecht had brought a working Altair 8800 with him which he demonstrated to the gathered enthusiasts, and there was much discussion as to what they might do with their computers once they got them working. Moore wrote up the meeting in the first edition of what was to become a monthly newsletter.

By April, membership had grown to 60 and meetings of the Homebrew Computer Club, as it was now called, were moved to Stanford University. It was at this meeting that Steve Dompier, one of the few members to have actually built an Altair 8800, demonstrated its musical capability. Having spent 30 hours constructing the kit (including six tracking down a broken connection), Dompier had run a test program and discovered that the process of executing the code generated electrical signals that were picked up by his transistor radio and turned into sounds that had a definite musical pitch. It took him a further eight hours to write the assembly code necessary to render a recognisable version of the Beatles song 'Fool on the Hill', which he demonstrated at the Club along with 'Daisy' (the tune sung by the dying HAL 9000 in the film *2001: A Space Odyssey*) as an encore (Dompier, 1975). As Felsenstein remembers:

> The radio began playing 'Fool on the Hill' and the tinny little tunes that you could hear were coming from the noise that the computer was generating being picked up by the radio. Everybody rose and applauded. I proposed that he receive the stripped Philips Screw Award for finding a use for something previously thought useless, but I think everybody was too busy applauding to even hear me. (Felsenstein, 1996)

At the same meeting another member demonstrated a TV Typewriter that he had constructed. The TV Typewriter had first appeared as a cover feature for *Radio-Electronics* magazine over a year earlier. Don Lancaster's article had described the construction of a computer terminal that used a low-cost keyboard and a radio frequency modulator to display 16 lines of 32 characters each on the screen of an ordinary domestic TV, all for around $120 (excluding the cost of the TV itself). Printed circuit boards for the project were available from Southwest Technical

Products in Texas. SWTPC went on to launch a more sophisticated version in January 1975 which was advertised on the page facing the Altair 8800 article in *Popular Electronics*.

Meanwhile, unable to meet demand, MITS had decided not to ship any peripherals until it had fulfilled all outstanding orders for the Altair 8800 computer itself. For those who already had a machine, the lack of memory expansion cards was particularly galling as the computer itself came with barely enough memory to run a few simple diagnostic programs. Furthermore, MITS had decided to use Dynamic RAM (DRAM) technology for its 4KB cards, and when these finally did arrive, the circuitry that the cards needs to keep the memory refreshed proved unreliable. Bob Marsh, who had attended the first meeting of the Homebrew Computer Club with Felsenstein, decided to do something about it, persuading his friend Gary Ingram that they should form a company and start selling their own expansion cards for the Altair. That April, the two of them formed Processor Technology and started advertising 4KB memory cards that used Static RAM (SRAM), so avoiding the need for additional circuitry and making them more reliable, at prices that undercut those of MITS.

The design of the Altair Bus – and indeed the Altair 8800 itself – was public knowledge, and Roberts had initially been happy for others to build expansion cards for his microcomputer as it would make it more useful and so increase sales. The August edition of the MITS newsletter *Computer Notes* made this clear, but with reservations:

> There is very little we can do or even want to do to prevent Altair 8800 customers from using these devices. The days of Big Brother Computer Company requiring Big Brother users to use Big Brother peripherals are over - period. However, this does not mean that we don't intend to be competitive. One obvious example of this is the structure of our software prices. (Bunnell, 1975, p. 2)

The price structure alluded to was that of Altair BASIC, as written by Gates and Allen. Those who had bought an Altair 8800 together with a 4KB memory card and a serial or parallel interface card from MITS, at a total cost of around $800 (over $1,000 assembled) could purchase Altair 4K BASIC for just $60: those who had bought their cards elsewhere had to pay $350. However Roberts was still not happy with the situation, and by the October edition of *Computer Notes* was referring to "parasite companies" who "must attack us for new business but at the same time they are dependent on our product for their survival."

Then, in December 1975, a company called IMS Associates announced the IMSAI 8080 computer. This was a mildly improved version of the Altair 8800 that used the same expansion bus design and was able to work with the same expansion cards. Almost overnight, companies like Processor Technology were no longer dependent on MITS for their survival.

Roberts reacted badly. The Computer Store had opened that July in Los Angeles, and by the end of the year computer shops were opening across the country. Roberts forbade MITS dealers, such as The Computer Store, from selling products that competed with the Altair range, which left independent outlets like Computer Mart in New York to sell IMSAI computers alongside expansion cards from Processor Technology and others (Veit, 2002). Roberts followed up by refusing to allow competitors to exhibit at the First World Altair Computer Convention, held in Albuquerque the following March, and was allegedly furious when he discovered that Processor Technology and others had taken rooms at the hotel where the convention was being held, and were using them to demonstrate their products (Levy, 2010, p. 230).

Roberts was also upset by what he saw as his loss of control of the Altair Bus. Initially, companies such as IMS and Processor Technology used the term 'Altair Bus' in their advertising material to refer to the expansion bus in the Altair computer. However it was on occasion referred to as the 'Altair/IMSAI Bus', and at some point it became more generally known as the 'Standard 100 Bus', or S-100 for short. Exactly how this came about is not clear, but it has been suggested that the term was introduced by the editor of an electronics magazine at the Personal Computing '76 show held in Atlantic City in August (Galletti, 2007). Responding to this, Ed Roberts' column in the November 1976 issue of *Computer Notes* contained the following, under the heading 'Charlatans, Rip-off Artists and Other Crooks':

> There is now an active attempt by a small group to steal the Altair bus. They are attempting to do this by changing the name so that they will not have to give recognition to MITS for its pioneering efforts in the small computer field. It is clear that the Altair bus is well established and the changing of the name to S-100 does not clarify or improve any situation for the user. It only helps the advertising copy of our competitors.
>
> The Altair bus was designed at least a year prior to the appearance of any of these competitors. The correct name for the Altair bus is simply the Altair bus, it

is not the Altair/IMSAI or the Altair/IMSAI/Polymorphic or the S-100, etc. Your help in stopping this sham will be appreciated, I hope you will identify the use of any name other than the Altair bus for what it is. (Roberts, 1976)

It was to prove a futile battle: at the West Coast Computer Faire in 1978, a variation of the S-100 bus was suggested to the Institute of Electrical and Electronic Engineers (IEEE) for adoption as a recognised standard, and in July 1979 a preliminary specification was published as IEEE 696. By this time it was in widespread use and many had forgotten that its roots lay in the Altair 8800.

Also becoming increasingly frustrated, although for a different reason, was Bill Gates. Anyone with a computer based around the Intel 8080 processor was a potential customer for Altair BASIC, which Micro-Soft supplied to MITS under contract. However the pricing policy adopted by MITS meant that a large proportion were being asked to fork out an exorbitant $350 for the privilege. This was at a time when few were accustomed to paying anything at all for software. The PCC and the Huntington Project had been based around the free exchange of software as an educational resource, while in the mainstream, companies like IBM and DEC tended to include software as part of the overall package. A few companies had been selling application software to mainframe and minicomputer customers since the 1950s, but the operating system for the IBM System/370 mainframe, for example, was in the public domain, and so freely available for use on compatible machines made by IBM's competitors (DeLamartar, 1986, p. 221).

Indeed there was a widespread belief amongst those who had come of age in the Summer of Love that business transactions were part of the capitalist culture that they wanted to transcend, and that creative work, such as music and software, should be freely shared for the greater good of all. There had been ugly scenes at the 1970 Isle of Wight music festival, for example, where many felt entry should be free, despite the obvious costs involved in setting up such an event. This belief is a common thread that runs through to today, manifesting in the Open Source movement (of which more later) and in repeated clashes with the music industry, first over home taping and later over Internet-based distribution services such as Napster.

So when MITS visited Palo Alto in June 1975, bringing with them an Altair 8800 connected to a teletype terminal and running Altair BASIC from a paper-tape reader, it was perhaps not surprising to find a Homebrew Computer Club member 'borrowing' a tape, duplicating it with the help of a friend with access to a tape-copying machine, and handing out copies of Altair BASIC at the next Club meeting free of charge (Levy, 2010, p. 231). And this was by no means an isolated incident, prompting Roberts to write in the October 1975 issue of *Computer Notes*, "Anyone who is using a stolen copy of MITS BASIC should identify himself for what he is, a thief." Gates himself followed up the following February with an 'Open Letter to Hobbyists' which was published in a number of newsletters, including *Computer Notes* and that of the Homebrew Computer Club:

> Almost a year ago, Paul Allen and myself, expecting the hobby market to expand, hired Monte Davidoff and developed Altair BASIC ... The feedback we have gotten from the hundreds of people who say they are using BASIC has all been positive. Two surprising things are apparent, however, 1) Most of these 'users' never bought BASIC (less than 10% of all Altair owners have bought BASIC), and 2) The amount of royalties we have received from sales to hobbyists makes the time spent on Altair BASIC worth less than $2 an hour.
>
> Why is this? As the majority of hobbyists must be aware, most of you steal your software. Hardware must be paid for, but software is something to share. Who cares if the people who worked on it get paid?
>
> Is this fair? One thing you don't do by stealing software is get back at MITS for some problem you may have had. MITS doesn't make money selling software. The royalty paid to us, the manual, the tape and the overhead make it a break-even operation. One thing you do do is prevent good software from being written. Who can afford to do professional work for nothing? What hobbyist can put 3-man years into programming, finding all bugs, documenting his product and distribute for free? The fact is, no one besides us has invested a lot of money in hobby software. We have written 6800 BASIC, and are writing 8080 APL and 6800 APL, but there is very little incentive to make this software available to hobbyists. Most directly, the thing you do is theft ... (Gates B. , 1976)

While Gates certainly had a point, the response from many was anger and outrage. Some argued that software should be in the public domain, pointing out that Gates and Allen had made extensive use of the PDP-10 at Harvard University, which had been financed using taxpayers' money. There was some truth in this and indeed Harvard did express concern when it discovered that its machine was being used for commercial gain. However the university did not have a policy that explicitly addressed such issues. Furthermore Gates and Allan did switch to a local timesharing service once the matter had been raised (Wallace & Erickson, 1992, pp. 82-83).

The response of Homebrew Computer Club member Jim Warren was on a different tack:

> There is a viable alternative to the problems raised by Bill Gates in his irate letter to computer hobbyists concerning 'ripping off' software. When software is free, or so inexpensive that it's easier to pay for it than to duplicate it, then it won't be 'stolen'.

He went on to describe two examples, before continuing:

> it is reasonable to expect that free and inexpensive software will become increasingly available to and through the hobbyists' community. This is true, in spite of the failure of such SHAREing in the business and industrial communities.
>
> 1. Hobbyists are developing home-grown hardware and software, just for the fun of it. Since it's 'fun' rather than 'work', they have shown a great willingness to share and distribute what they develop. This is not an unknown phenomenon. It is the usual practice in most other hobby environments, and is certainly true in the academic environment.
>
> 2. As with the industrial mini and micro markets, hobbyists have learned to be wary of purchasing hardware from manufacturers who provide no software support. Through common sense, and by observing Mr. Gates' experience, those who wish to sell software for significant sums of money must realize that there is only one group that can practically be expected to pay for it: the hardware manufacturers. They need it to enhance their products in a highly competitive marketplace.
>
> 3. Concerning quality: A significant minority of computer hobbyists are also experienced computer professionals. It's their (our) play as well as work. The competency level is more than sufficient for the design and implementation of excellent systems software.
>
> 4. Finally, the approach used in producing the Tiny BASICs will be continued and expanded, a sort of modified Chief Programmer Team approach: An experienced pro does the overall design and outlines the implementation strategy (via the Journal and other hobbyist publications). Following those directions, the more experienced amateurs do the necessary hack-work (exciting to them, but drudgery for the 'old pro'). Since it is a symbiotic effort, the implementers are almost certain to share their work with the designers, and hence, with the larger community of home computer users.
>
> It's amazing how much 'good stuff' becomes available when the producers think of their labor as 'play' instead of 'work'. All who wish to do so are invited to join with the publishers of *Dr. Dobb's Journal* in the pursuit of realizable fantasies. (Warren, 1976)

The principal example used by Warren was that of Tiny BASIC. Frustrated by the pricing structure imposed by MITS on Altair BASIC, Bob Albrecht had persuaded Dennis Allison (a computer science lecturer and a co-founder of the People's Computer Company) to write an article for the PCC newsletter called 'Build your own BASIC', proposing a syntax for a BASIC interpreter that would be able to run in the limited memory space available in microcomputers such as the Altair 8800. Within weeks of publication he was receiving letters suggesting improvements, and even an implementation in machine code for the 8080 processor.

Realising that he could not do justice to this level of interest in the existing newsletter, Albrecht decided to create a new magazine, and persuaded Warren to become editor. The first issue of *Dr. Dobb's Journal of Tiny BASIC Calisthenics & Orthodontia* (quickly shortened to the *Dr. Dobb's Journal* referred to above) appeared in January 1976, with the opening page proclaiming:

> This newsletter is meant to be a sharing experience, intended to disseminate FREE software. It's OK to charge a few bucks for tape cassettes or paper tape or otherwise recover the cost of sharing. But please make documentation essentially free, including annotated source code.

Tiny BASIC did not support as many instructions as Altair BASIC, but interpreters were available for the Intel 8080 and Motorola's 6800 processor (see below) on cassette tape for just five dollars, or for free if you were prepared to type in the necessary machine code yourself.

Warren's letter was proposing an academic approach to software development, with source code freely available for anyone to enhance. Although individual contributions would generally be recognised, usually through 'comment' statements embedded in the code itself, no one would be allowed to claim ownership of the code. These days such an approach is referred to as 'open source'. Gates, on the other hand, was making the case for software as a business, with programmers owning and being financially recompensed for their work. Warren felt that such an approach couldn't work as hobbyists wouldn't be able to afford the high cost involved, but this was a new world, and the old rules were rapidly becoming irrelevant.

The debate continues today (and is covered in our chapter 'Geeks bearing gifts' starting on page 213), but at the root of this particular spat was the pricing policy adopted by Roberts. Furthermore, Roberts made it clear that he was not interested in licensing Altair BASIC to anyone else, realising that to do so would be to lose the competitive edge that it gave his products. Then, in early 1977, Gates learned that Roberts was considering an offer from Pertec, who supplied MITS with disk drives, to buy the company outright. Pertec assumed the deal would include exclusive rights to Altair BASIC through the contract that MITS had signed with Microsoft, but Gates was determined to break free. Following consultations with his father and a local attorney, he wrote to Roberts terminating the agreement on the basis that MITS had failed to "use its best efforts to license, promote, and commercialize the Program (BASIC)", as was required by the contract. Gates' case was strengthened by a letter that he had received from Pertec explicitly stating that the company did not intend to allow BASIC to be licensed to third parties. The case went to arbitration and was eventually resolved in Gates' favour (Wallace & Erickson, 1992, pp. 111-116).

Having severed ties with both MITS and Pertec, Gates and Allen had no reason to stay in Albuquerque and so, at the beginning of 1979, Microsoft moved to Seattle where its headquarters remain today. Roberts used part of the proceeds of the sale to buy a farm in Georgia, before training to become a doctor in the small town of Cochran. Both Gates and Allen were in contact with Roberts when he fell ill, and on his death in 2010, they paid tribute:

> Ed was willing to take a chance on us — two young guys interested in computers long before they were commonplace — and we have always been grateful to him. (Lohr, 2010)

Pertec produced a number of microcomputers, none particularly successful, before being purchased by its European distributor Triumph-Adler in 1980.

Neither Intel nor MITS had the market to themselves for very long. Motorola launched its 6800 processor in August 1974, just months after Intel's 8080 had appeared. This was very similar to the Intel device, but not directly compatible. Like the 8080, the 6800 was part of a family that included memory chips, peripheral interface chips and – in this case – a modem chip which supported 600 bits per second communication over a telephone line. As with the Intel 8080, the price at launch for a single chip was $360, although within a year that figure had halved.

South West Technical Products followed up its earlier TV Typewriter kit with a microcomputer kit based on Motorola's 6800 in November 1975, at the same time as MITS launched the Altair 680 as a project in *Popular Electronics*. Neither was to prove as successful as the Altair 8800, but the SWTP 6800 is notable for its inclusion of SWTBUG, the company's own version of the MIKBUG software that Motorola programmed into the 6830 ROM chip that was part of the 6800 family. SWTBUG was a machine code monitor – in other words a very rudimentary operating system – that allowed users to perform simple operations like loading and saving programs from cassette or paper tape, or communicating with a terminal, without having to laboriously key in a whole series of machine code instructions each time. Instead the instructions were stored in ROM, ready to be triggered by a single command.

Following the release of the 6800, Chuck Peddle, Bill Mensch and a number of other Motorola engineers who had worked on the chip became aware that, while there was certainly a great deal of interest in the device, many potential customers were put off by the price. They became convinced that a device costing just $25 could be viable because it would sell in far greater quantities. Motorola management made it clear that the company was not interested, so in early 1975, Peddle and his colleagues left Motorola to join MOS Technology, a small fabrication plant in Pennsylvania that was prepared to take the gamble. MOS Technology had developed a process that greatly reduced the proportion of defective chips that came off the production line, which helped to keep costs down.

MOS Technology's first microprocessor was the 6501. This was very similar to Motorola's 6800 and was driven by the same 1MHz clock. However it could execute instructions considerably faster, thanks to a rudimentary form of instruction pipeline which meant that one part of the processor could be loading data for the next instruction while another part was still working on the first. Furthermore the pin arrangement of the two chips was identical, a rather cheeky move which meant that the 6501 could be plugged directly into a motherboard designed for a 6800.

Unimpressed, Motorola sued for breach of copyright. MOS Technology promptly withdrew the 6501 and replaced it in September 1975 with the 6502, launched at the Western Electronics Show and Convention (WESCON75) in San Francisco. The 6502 was essentially identical to the 6501 but with a different pin arrangement, and was priced at $25 (Matthews, 2003). Just one month later, Motorola reduced the price of the 6800 from $175 to $69. The lawsuit dragged on for several years and was finally settled with a payment to Motorola.

One person who was to make good use of the 6502 was Steve Wozniak, a college dropout from San Jose who had found a job as an engineer with Hewlett-Packard, working on electronic calculators. Wozniak (or 'Woz' as he is known to his friends) was obsessed with electronics, and saw it as a personal challenge to implement a circuit using as few components as possible. This was partly to save money but also because, as he put it: "I wanted to put chips together like an artist, better than anyone else could and in a way that would be the absolute most usable by humans" (Wozniak & Smith, 2006, p. 18).

Wozniak was a close friend of Steve Jobs, a young hippy who had immersed himself in the West Coast scene; smoking marijuana, taking LSD and studying Zen Buddhism. After leaving college, Jobs had started work at the video game company Atari, set up by Nolan Bushnell a couple of years earlier, although he took time out in the summer of 1974 for the traditional pilgrimage to India (Isaacson, 2011, p. 42). Jobs was very different to the shy and unassuming Wozniak, being five years younger and more at home hustling for business, but they were drawn together by a mutual fascination with anything electronic.

Atari's first product had been *Pong*, an arcade game developed by Al Alcorn that let you play tennis for an allotted time once you put a quarter in the slot. The *Pong* machine itself consisted of a board of discrete logic chips connected to a black-and-white TV set and two controls for moving the bats up and down the screen. The circuitry was hard-wired to play the game – there was no software, in other words – and the whole lot was housed in a wooden cabinet, together with the coin slot. It proved hugely successful; so much so that, when the prototype broke down just days after installation at Andy Capp's Tavern in Sunnyvale, California in December 1972, the problem was tracked down to the coin mechanism having shorted out because it had become jammed with quarters.

Bushnell had met Wozniak through Jobs and been impressed by his skills, so he asked Jobs to build a prototype for a new arcade game called *Breakout*, knowing that Wozniak would become involved. The game involved moving a paddle across the bottom of the screen so as to keep a ball bouncing against the layers of bricks that made up the top of the screen. Hitting a brick made the brick disappear, until eventually the ball could break through the wall and you won the game. If you missed the ball, you lost. One of the attractions of *Breakout* was that it only required a single player, unlike *Pong* which needed two people to play.

As the two friends worked together on this project, Jobs mentioned to Wozniak that Atari was planning at some point to replace some of the discrete circuitry with a microprocessor. Then, a colleague from Hewlett-Packard persuaded Wozniak to attend that first meeting of the Homebrew Computer Club, held in Gordon French's garage in March 1975, where he encountered the Altair 8800. He was too shy to say anything while at the meeting, but on returning home, the realisation came to him that he could use the knowledge he had gained working on *Breakout* to build something far better, and cheaper.

Wozniak was not thinking of a commercial product, but rather a computer that he could build for himself and show the people he had met at the Homebrew meeting. He wanted it to have a screen and a keyboard, rather than rows of lights and switches. He also decided to include a ROM chip that would store a 'monitor' program designed to run automatically when the device was turned on, so that it would know straight away how to respond to key presses and use the display. With the Altair 8800, such a program had to be entered manually as binary code using the front panel switches before the computer would do anything. And finally, he wanted it to be affordable.

Initially Wozniak intended to use the Motorola 6800, but when he heard about the 6502, he drove up to WESCON75 and bought a few directly from the MOS Technology booth. He decided to employ dynamic rather than static memory chips, allowing him to offer greater storage capacity, and he overcame the refresh problem by ingeniously 'borrowing' the video display circuitry for a few microseconds between each scan so that it could refresh the dynamic memory. There was a version of Tiny BASIC available for the 6502, but Wozniak wanted something a bit more comprehensive, so he wrote his own. It became known as Integer BASIC because it could only work with variables that were either integers or character strings.

The end result was a printed circuit board that had room for 8KB of memory, a slot for an expansion card, and connections to a typewriter-style keyboard and either a TV or a video monitor. The display was very slow by modern standards, but considerably faster than the teletype terminals commonly used with the Altair 8800, and displayed 24 rows of 40 alphanumeric characters each line.

Wozniak demonstrated his device at Homebrew Computer Club meetings and was happy to hand out drawings so that others could build their own. There was considerable interest, but few had the time or expertise to follow in Wozniak's footsteps, so Jobs suggested that they pool resources and form a company that could sell the printed circuit boards ready-made. Jobs reckoned they could produce the boards for $20 each and sell them for $40, making $20 in profit. Setting up the screen from which the boards would be printed cost $1,000, so they needed to sell 50 boards if they were to recoup costs. On 1 April 1976, the two of them formed the Apple Computer Company – the name was suggested by Jobs, and although Wozniak was a little worried that Apple Corps, owned by The Beatles, might object, they couldn't think of anything better (Wozniak & Smith, 2006, p. 173). In the event, Apple Corps did sue Apple Computer for trademark violations just two years later. The suit was settled in 1981 with a payment to Apple Corps of $80,000 and an agreement that Apple Computer would stay out of the music business, something which would cause further problems with the launch of iTunes several decades later (Salkever, 2004).

Meanwhile, Paul Terrell had opened the Byte Shop in Mountain View, California, and was looking for new products to sell. He had seen one of Wozniak's demonstrations and expressed his interest to Jobs. However he argued that it would be much easier to sell ready-made computers than naked printed circuit boards, so he agreed to buy 50 units at $500 a time, payable as each was delivered, but only if they were pre-assembled. Apple had already arranged the boards, and Jobs managed to negotiate 30 days' credit with a component supplier which meant that they could pay for the components after they had delivered each computer, so obviating the need to raise any money up front. They had a deal.

Terrell realised that the package would be even more attractive if it came with some sort of storage device: as it stood, you had to key in the machine code that made up the Integer BASIC interpreter before you could use it, which could take an hour or more. He therefore insisted that Wozniak build a cassette interface card that could plug into the expansion slot and allow programs to be written and read from an audio cassette player. The assembled boards went on sale as the Apple-1 that July at a price of $666.66 with 4KB of memory ($500 plus a one third mark-up for Terrell). The cassette interface was an extra $75, which included a copy of Integer BASIC on cassette tape.

Finally, Terrell suggested that they hide the complexities of the electronic circuitry from view. They weren't producing enough units to merit commissioning a plastic case, so Terrell arranged for a case made of koa, a wood native to Polynesia, to be sold as an optional extra.

In the end just 200 were made, of which all but 25 were sold over the next nine or ten months (Williams & Moore, 1984). By contrast, MITS had sold over 10,000 of its Altair units, but the Apple-1 set a precedent, establishing that microcomputers should have a typewriter-style keyboard rather than a row of switches; should display results on an ordinary TV set or video monitor rather than a row of light bulbs; and use conventional audio cassettes for storage. Furthermore, as its name suggested, Wozniak was already working on its successor.

4.   Fun and games

The Apple II was the first of a new generation of computers powerful enough to appeal to a much wider audience – an audience interested in what a computer could do, rather than in the electronics that made it possible. This was not a printed circuit board and a bunch of chips that you had to put together yourself, or a box into which you had to enter arcane code before it would even recognise that you were there: it was something you could unpack, plug in, turn on and start using straight away.

The machine which Wozniak demonstrated at a Homebrew Computer Club meeting towards the end of 1976 used the same 6502 processor as its predecessor, the Apple-1. However it came with 8KB of ROM which contained not only the monitor program but also an implementation of Wozniak's BASIC, so that users could start writing BASIC programs straight away, without having to first load the interpreter from cassette tape. The board had room for three banks of eight memory chips, which if completely filled with the 4Kbit chips common at the time would provide 12KB of RAM for the user's programs. Alternatively (and at rather greater expense) they could be replaced with the new 16Kbit chips that were becoming available, giving you a whopping 48KB of RAM.

But what really made the Apple II stand out was that it could display colour on a conventional TV screen. In addition to a basic text mode showing 24 rows of 40 characters, it could display 48 rows of 40 cells, each in any of 15 colours (technically 16 but the two shades of grey were effectively identical). There was also a 'high resolution' mode that could display 280 by 192 pixels in four colours (black, white, violet or green). Just the video data required to display in this mode took up 8K of RAM, which in practice meant that you needed at least 12K installed if there was to be any space left over for your own programs.

The Apple II was a big hit at the First West Coast Computer Faire, held in San Francisco the following April. Organised by *Dr. Dobb's* editor Jim Warren, this event has been described as the Woodstock of the computer world (Levy, 2010, p. 273). Over a period of three days, more than 12,000 almost exclusively male enthusiasts converged on the Brooks Auditorium to exchange ideas, brush shoulders with their heroes and find out what the 175 exhibitors had on show. Those exhibiting included Intel, Processor Technology, SWTPC, MITS and IMSAI, as well as the People's Computer Company, the new *Byte* and *Creative Computing* magazines, and various retailers, component suppliers and book publishers. Ted Nelson, who delivered the keynote speech, captured the significance of the event with typical prescience:

> Here we are at the brink of a new world. Small computers are about to remake our society, and you know it … The little computers are here, you can buy them on your plastic charge card, and the available accessories include disk storage, graphic displays, interactive games, programmable turtles that draw pictures on butcher paper, and goodness knows what else. Here we have all the makings of a fad, it is fast blossoming into a cult, and soon it will mature into a full-blown consumer market. (Levy, 2010, p. 274)

Apple had secured a double booth near the entrance, where they programmed the new Apple II to display an ever-changing kaleidoscope of full colour graphics on a giant projection TV display. However they proved to be only one of many attractions. Another big hit at the Faire was the Commodore PET, which Commodore had announced at the Winter Consumer Electronics Show held in Chicago a few months earlier.

Jobs and Wozniak had, in fact, demonstrated a prototype of the Apple II to Chuck Peddle shortly after he had joined Commodore, and also to Job's old boss Al Alcorn at Atari, hoping that one of them might be interested in helping to get the Apple II into production. Jobs somewhat shocked Commodore by suggesting that they might want to offer "a few hundred thousand dollars" for the Apple II, and put Jobs and Wozniak in charge of the development programme. Commodore declined, opting instead to bring out its own machine. Atari, still heavily involved in the video games market, also turned them down, but in the process put them in contact with Mike Markkula, who had retired from Intel at the age of 32 having made millions from stock options. Markkula joined Apple as its third employee and part-owner, bringing with him enough finance for them to go it alone.

Commodore had been founded by Jack Tramiel in Toronto, Canada during the 1950s, initially manufacturing typewriters and adding machines. The name of the company had come to him while sitting in the back of a taxi cab. He had served in the US Army and wanted "something strong" like General or Admiral, but those were already taken, and then he noticed that the car in front of him was an Opel

Commodore (Commodore History, 1989).

Like the Altair manufacturer MITS, Commodore had moved into the lucrative electronic calculator market in the early 1970s but found itself unable to compete when the chip manufacturers started building them too. The Canadian investor Irving Gould had already bought a substantial stake in Commodore during the 1960s, helping the company to recover after one of its backers was indicted for fraud. In 1975 Gould again came to the rescue by personally guaranteeing a $3 million loan which enabled Commodore to make a series of strategic acquisitions.

Perhaps the most significant of these acquisitions was the purchase of MOS Technology in 1976, which brought with it the 6502 processor and Chuck Peddle, responsible for much of the processor's design and full of ideas as to what could be done with it. Peddle persuaded Tramiel that the future lay in microcomputers, and Tramiel responded by giving Peddle a free hand. The result was the Commodore PET: the name officially stood for Personal Electronic Transactor, but actually owed more to the Pet Rocks that had become a fad the previous Christmas (stones that were packaged as the 'perfect pet' and sold for $3.95 complete with a 32-page training manual).

Technically the Commodore PET was similar to the Apple II in that it used the same processor running with the same 1MHz clock, and came with either 4K or 8K of RAM. It also included 14K of ROM, of which 8K was taken up with a version of BASIC written for Commodore by Microsoft. However the PET did not compare well with the Apple II with regard to graphics, only being capable of displaying 25 lines of 40 monochrome characters, although these did include a range of graphic symbols made up of lines and blocks, together with the heart, diamond, club and spade symbols found on playing cards.

So for the real enthusiast the PET was not as exciting as the Apple II, but where it did score was on price, with the 4K version initially advertised at a mail order price of just $495, and the 8K version at $795. This was for a unit that included a nine-inch video screen and a cassette drive built in to the case: by contrast, the Apple II cost $1,298 for just the keyboard and the electronics in a plastic box (equivalent to around $4,800 today). Once the PET started shipping, Commodore increased the price to $595, which had the knock-on effect of increasing demand for the 8K version, now only $200 more.

Then there was the TRS-80, announced in August 1977 by the Radio Shack division of Tandy Corporation (hence 'TRS'). This came with 4K of RAM and 4K of ROM which contained the software necessary to interface with a cassette recorder, together with Radio Shack's Level 1 BASIC interpreter (actually a version of the open source Tiny BASIC). Like the PET, the display was fairly primitive, capable of displaying 16 lines of 64 characters of monochrome text. However the matrix into which the characters were drawn could be further divided into six squares (two across and three down) which could be individually turned on or off, so supporting graphic displays with an effective resolution of 128 by 48 points.

For the TRS-80, Tandy chose not to use the 6502 processor adopted by the Apple II and the Commodore PET, but instead plumped for the Z80 processor which Zilog (the company set up by Frederico Faggin after leaving Intel in 1974) had launched the year before. This was essentially an improved version of the 8080, answering to the same instruction set but coming with duplicate banks of registers which allowed it to respond more quickly to external interrupts triggered when, for example, the user pressed a key on the computer's keyboard. The Z80 had the circuitry necessary to refresh dynamic memory built in, which made it much easier to work with dynamic RAM chips, and it needed just a single power supply, where the 8080 required three separate voltages to function. Furthermore, it could work with a 2.5MHz clock which made it inherently faster than its competitors.

The TRS-80 microcomputer was competitively priced, costing just under $600 complete with a Radio Shack video display and cassette recorder, or $400 for the keyboard unit by itself, but its big advantage was the Radio Shack branding. Radio Shack had long been a household name, selling electronic goods and components to domestic consumers and enthusiasts through its network of 6,000 stores. Within the first month, Radio Shack sold 10,000 units, and by the end of the following year, *Byte* magazine was reporting sales of 100,000 units. This was four times as many as the Commodore PET and five times as many as the Apple II. By contrast, IMSAI had managed sales of just 5,000 units, and the MITS Altair even less at 3,000 units (Libes, 1979).

With sales figures like these, the microcomputer was beginning to attract the attention of the general public – ordinary people who weren't interested in technology for its own sake, but did want to find out why anyone would want to own such a thing. On their part, the companies involved realised that they needed to expand the appeal of their new devices beyond the geeks who were happy to spend all night typing machine code. They needed to promote the microcomputer as a useful addition to not only business, but also to family life. Accordingly, the first Apple II advert began:

> Clear the kitchen table. Bring in the color TV. Plug in your new Apple II, and connect any standard cassette recorder/player. Now you're ready for an evening of discovery in the new world of personal computers. Only Apple II makes it that easy. (Stengel)

Radio Shack was claiming for the TRS-80:

> The Family Christmas Gift That's Functional, Fun and Educational! (Stengel)

As always, the reality was rather different from the advertising copy. Both machines had to be placed at least a couple of feet away from the television in order to minimise the electrical interference they generated, while the "standard cassette recorder" proved unreliable as a storage device, requiring patient adjustment of tone and volume before the computer would recognise that there was any data in the noise coming off the tape.

Furthermore, there was little software available. Radio Shack did introduce five software packages for the TRS-80 which included a 'kitchen' program for recording and displaying recipes, a personal finance program for balancing your cheque book, an education program for testing basic mathematical skills, and a payroll package that could handle up to 15 employees (more if you installed additional memory). However these were of little more than novelty value. What were needed to excite particularly the younger members of the family were games.

By the time these machines were appearing, coin-operated electronic game consoles had become commonplace in bars, foyers and gaming arcades everywhere. Even in the most remote of public houses, bemused regulars were getting used to the unearthly whistles and burps that emanated from these consoles as groups of youngsters fed them with a seemingly endless stream of coins. It appeared that little could be done, as the revenue they generated often exceeded that coming from behind the bar. However, despite Atari's early success with *Pong* and *Breakout*, most of the games being played originated in Japan.

Atari had opened an office in Tokyo in 1973 but had not been able to penetrate the Japanese market, so Bushnell had sold Atari's Japanese business to Nakamura Manufacturing for a reported half million dollars, granting Namco (as the company was more commonly known) exclusive rights over Atari products in Japan for 10 years (Current, 2011). Namco went on to develop its own arcade games, the most successful being *Pac-Man*, released in 1980 and developed by Tōru Iwatani in an attempt to create something that would appeal to women as well as men (there was a fad amongst young Japanese women at the time for 'fun' foods which gave him the idea of a game based around eating).

Perhaps the most successful of all the coin-operated games to come out of Japan was *Space Invaders*, released in 1978 by Taito Corporation and licensed by Midway for production in the United States. Designer Tomohiro Nishikado was in part inspired by Atari's *Breakout*, but also influenced by the imminent release of the film *Star Wars* and the octopus-like aliens of the 1950 movie *War of the Worlds*. The game became so popular that it caused a shortage of 100 yen coins and brought in hundreds of millions of dollars in its first year.

Atari did release a number of successful coin-operated games, including *Lunar Lander* and the highly popular *Asteroids* in 1979, together with *Battlezone* and *Missile Command* in 1980. However, following his original success with *Pong* back in 1972, Bushnell had gone on to design a domestic version that could be played through a conventional TV set. The result was the Home Pong console, sold through the Sears and Roebuck mail order catalogue in 1975. Sears initially ordered 50,000 units, but later increased this to 150,000 units in order to meet Christmas demand. Atari seemed to be on to a winner, but by the following year, and in the face of stiff competition from the likes of Coleco and Magnavox, it was obvious that the investment required to design a dedicated games console, using standard logic chips, did not make sense. Far better to develop a general-purpose games console that could be programmed to play any number of games.

Atari was already investigating the viability of a console in which the games would be programmed into plug-in cartridges that could be sold separately, but had been put off by the high cost of the microprocessors necessary to make this a reality. However this all changed in September 1975, when two members of the development team met up with Chuck Peddle at WESCON75 in San Francisco, where he was promoting MOS Technologies' new low-cost 6502 processor, and negotiated a deal that would allow Atari to buy suitable quantities of an even cheaper version of the chip, called the 6507, at just $8 each. The result was the Atari Video Computer System (VCS), which was eventually launched in October 1977 at $199 (about twice the price of a modern Xbox or PlayStation, allowing for inflation) and sold through Sears and Roebuck as the Sears Video Arcade. The price included two joysticks and a plug-in cartridge containing the code necessary to play *Combat*, a two-player game that supported a number of combat scenarios featuring tanks, biplanes and jets. Eight other cartridges were available at launch including *Star Ship, Street Racer, Air-Sea Battle* and, for those concerned about the device's educational value, *Basic Math*.

Despite its 6507 processor, the VCS was rather different to a conventional microcomputer, comprising just four chips that included 128 bytes of RAM and a dedicated display and sound chip called the Television Interface Adaptor (TIA), which was capable of displaying an unprecedented 128 different colours. Games cartridges could contain up to 4K of ROM in which the code for the game was stored, although later games used bank switching to allow access to larger amounts of memory. Cartridges sold for around $20 to $30 each.

So by the time the Apple II, the Commodore PET and the TRS-80 became available, modern families were becoming accustomed to the idea of an electronic gadget that could play a variety of arcade-style games on a domestic TV set. As Wozniak himself admitted:

> A lot of features of the Apple II went in because I had designed *Breakout* for Atari. I had designed it in hardware. I wanted to write it in software now. So that was the reason that color was added in first – so that games could be programmed. I sat down one night and tried to put it into BASIC. Fortunately I had written the BASIC myself, so I just burned some new ROMs with line drawing commands, color changing commands, and various BASIC commands that would plot in color. I got this ball bouncing around, and I said, 'Well it needs sound', and I had to add a speaker to the Apple II. It wasn't planned, it was just accidental ... obviously you need paddles, so I had to scratch my head and design a simple minimum-chip paddle circuit, and put on some paddles. So a lot of these features that really made the Apple II stand out in its day came from a game, and the fun features that were built in were only to do one pet project, which was to program a BASIC version of *Breakout* and show it off at the [Homebrew] club. (Connick, 1986, p. 24)

These microcomputers were a great deal more expensive than an Atari games console, and considerably less reliable. Plug a cartridge into an Atari VCS and you could be playing the game in a matter of seconds, but loading a new program from audio cassette into an Apple II or TRS-80 could take several minutes, or even several attempts if you hadn't got the controls set just right.

On the other hand, audio cassettes were mass market items that didn't require specialist assembly. All you needed to start a games company was a microcomputer, an understanding of the BASIC programming language, a cassette recorder and a good supply of blank cassettes. If you were targeting the Apple II, and had the necessary skills, then you could write even better games using machine code, and Apple was careful to supply detailed documentation and programming tools to make it easier. By the late 1970s there were dozens of such companies, many little more than a couple of guys working from home, selling computer games on audio cassette through specialist shops, computer clubs or small adverts in specialist magazines such as *Byte* or *Creative Computing*, often for less than $10 each.

Initially these games offered little more than the BASIC programs listed in such magazines, or in books such as *101 BASIC Computer Games* or *What to Do After You Hit Return*. Early titles included *Kingdom, Luna Lander* and *Star Trek*, together with implementations of popular card and board games. However playable versions of arcade classics soon appeared, particularly on the Apple II with its colour display. These included a version of *Space Invaders* called *Super Invader* which was voted 'Most Popular Program of 1978-1980' by the readers of the Apple II magazine *Softalk* (Weyhrich), and *Asteroids* which could trace its roots to the 1960s classic *Spacewar*.

More ambitious titles started to appear as programmers got to grips with these new machines. Bruce Artwick had been writing a program in assembler language for Motorola's 6800 processor that would simulate the experience of flying an aircraft, based on work he had been doing as a student at the Digital Computer and the Aviation Research Labs at the University of Illinois. In the process he wrote a number of magazine articles which stimulated interest, and in 1977 he started a company called subLOGIC to distribute the code on audio cassette. The company went on to release *A2-FS1 Flight Simulator* for the Apple II in January 1980, followed a couple of months later by a version for the TRS-80 called *T80-FS1*.

The program displayed the instrument panel of a Cessna light aircraft in the lower half of the screen, including altimeter and speedometer, while the top half showed the view through the windshield in a simple monochrome 'wire-frame' format that refreshed just three times a second. Players could attempt to take off, fly around and land by controlling throttle, rudder and elevator, and there were even controls for declaring war, dropping bombs and shooting guns. It proved extremely popular, despite the jittery display, and is still available today, albeit in a considerably more sophisticated form, as *Microsoft Flight Simulator*.

Another genre of computer game that flourished on these early microcomputers was the text adventure, largely because it did not need colourful high-speed graphics but instead, like a good book, relied on the imagination of the player. Indeed, at their best, such games could better be described as interactive novels.

Adventure games predate microcomputers, first appearing in the early 1970s with the likes of *Hunt the Wumpus*, written by Gregory Yob in BASIC and published in a 1973 edition of the People's Computer Company newsletter. There were already games that involved hunting for something that was hidden in a grid, but Yob realised that a computer could support more interesting topologies which would make such games more challenging. *Hunt the Wumpus* creates a labyrinth of 20 rooms, each connected to three others in a three-dimensional geometry mirroring that of a dodecahedron. One of the rooms contains a Wumpus that will eat anyone entering the room, while some of the other rooms contain bottomless pits. Thankfully you can tell if you are adjacent to a Wumpus by its stench, or a bottomless pit by a breeze. Your job is to negotiate the labyrinth and kill the Wumpus before you are eaten or fall down one of the bottomless pits.

This was followed by *Adventure*, a much more sophisticated example of the genre originally written by Will Crowther in 1976 while working in Boston. Crowther wrote the code in FORTRAN to run on a PDP-10 mainframe, and the game was based in part on his explorations of the Colossal Cave complex in Kentucky. The underlying structure was similar to *Hunt the Wumpus*, in that the player was presented with a maze of interconnected locations. However each location had a detailed description, and the program understood a far wider range of instructions.

As was common in those days, the program quickly found its way onto computers on campuses across the continent. It was picked up by Donald Woods, a graduate working at Stanford Artificial Intelligence Laboratory (SAIL) in California, who contacted Crowther for a copy of the source code, which he expanded by adding new puzzles and more complex elements such as an underground volcano and a vending machine that dispensed batteries for your headlamp (Jerz, 2007).

The first person to implement a text adventure on a microcomputer was Scott Adams, who came across *Adventure* on the company mainframe while working in Florida for telecommunications company Stromberg-Carlson. Adams owned a

TRS-80 and, realising that he could never fit something as extensive into its 16K of memory, decided to write his own version. The result was *Adventureland* which involved searching "an enchanted world" for 13 artefacts. Once finished, Adams took out a small advertisement in the magazine *Softside* and started selling the game mail-order at $14.95 a time, using his TRS-80 to copy the code onto blank audio cassettes. The breakthrough came when the manager of a Radio Shack in Chicago ordered 50 copies. As Scott Adams recalls, "it took a while to create those 50 tapes one by one on my TRS-80!" (alistairw, 2006).

The Crowther and Woods program arrived at MIT around May 1977 where it inspired a group of hackers to write something similar called *Zork*. This was initially written for the laboratory's PDP-10 mainframe, but they went on to create a version for the TRS-80 which was completed by early 1980, followed at the end of the year with a version for the Apple II. In the process, they had to abandon half of the game locations in order to fit the code into the memory available, and design a new computer language called ZIL (Zork Implementation Language) for the purpose. They formed the company Infocom as a vehicle for their programming efforts, although rather than sell direct, they decided to use one of the first software publishing companies to distribute the program, namely Personal Software (which made its name distributing the VisiCalc spreadsheet program). The Apple II version of *Zork* proved particularly popular, with Personal Software selling 6,000 copies in the first eight months. The follow-up, *Zork II*, was initially intended simply as an implementation of the original game locations that had been left out, but designer David Lebling kept coming up with new puzzles and the game eventually ran to three instalments (Anderson & Galley, 1985).

5.   Extraordinary fellows

Such games were increasing the demand for these early microcomputers, but the fact remained that the hardware needed to play them was extremely expensive. In May 1979, an article in the *Financial Times* predicted, "Personal computers will become steadily cheaper and their price could drop to around £100 within five years." At the time, the Tandy TRS-80 was available in the UK for just under £500 (equivalent to nearly £2,000 today) while £450 might just buy you a Commodore PET. An Apple II, even in its most basic configuration, cost well over £1,000. However the price of microprocessor and memory chips was coming down, and some people weren't prepared to wait.

The Cambridge-based entrepreneur Clive Sinclair had already built up a reputation for delivering technical gadgetry at very low prices through his company Sinclair Radionics. Early examples include the Sinclair Micro Amplifier, which he made available in 1962 for just £1.50 (equivalent to about £24 today), and the Sinclair Slimline pocket radio a year later for £2.50, both sold on the basis that you could "build it in a couple of hours" from the collection of electronic components provided. They were followed in quick succession by the "world's smallest radio" in the Sinclair Micromatic; the IC-10 integrated circuit amplifier in 1969; and the Executive calculator in 1972, which won a Design Centre award. All were characterised by their small size and low price, and were eagerly snapped up by geeky teenagers across the country. As a result, Sinclair was soon being referred to as 'Uncle Clive', despite many of his products failing to live up to expectations. The Black Watch, for example, (a digital watch launched in 1975 as a kit costing £17.95) proved extremely fiddly to assemble and, if you did eventually succeed in getting it to work, notoriously sensitive to static electricity.

As a result of production problems with the Black Watch, the company started to lose serious money. Nevertheless Sinclair continued to finance research into his pet projects, namely the Microvision pocket television and his abiding obsession, the electric car. In June 1976, having been rejected by private sources, he approached the National Enterprise Board for investment, and a couple of months later agreed to give the NEB a 43 per cent share of the company in exchange for

£650,000. This proved sufficient to launch the Microvision, but not to maintain production. Accordingly the NEB invested a further £1.65m in exchange for an additional 30 per cent of the company, which left Sinclair a minority shareholder (Adamson & Kennedy, 1986, pp. 40-42).

In contrast to the United States, Britain in the 1960s and 1970s could perhaps best be described as a socialist democracy, still recovering from the devastating impact of the Second World War. The National Research Development Corporation (NRDC) had been set up in 1948 by an earlier Labour government in an effort to facilitate the commercialisation of the many publicly funded projects that had come out of the war effort. It was joined by the NEB in 1975, created under Prime Minister Harold Wilson's newly returned Labour government, by which time the remit had expanded to include public investment in private companies that were deemed important to the British economy – particularly those competing against the Japanese and the Americans in the "white heat" of the technological revolution to which Wilson had referred in his 1963 speech to the Labour Party conference.

Not one to limit his options, Sinclair had taken the precaution back in 1973 of buying an off-the-shelf company called Ablesdeal Ltd, changing its name first to Sinclair Instruments and then, in July 1977, as his control of Sinclair Radionics slipped away, to Science of Cambridge. At around this time, he also persuaded his colleague Chris Curry to leave Sinclair Radionics and start work building up his new company. Sinclair's long-term goals remained the pocket TV and the electric car, but in the meantime Curry concentrated on generating cash. Accordingly, the company's first product was the Wrist Calculator, released in February 1977 which, although no easier to assemble than the Black Watch, proved surprisingly successful. Then Curry was approached by Ian Williamson, a Cambridge undergraduate with an interesting proposition.

Like many of his peers, Williamson had become fascinated with microcomputers but disillusioned by their high prices. He thought that it should be possible to put together something usable that could be sold at a much lower price, and had managed to build a very basic microcomputer around the SC/MP (short for Single Chip Micro Processor and usually pronounced "Scamp"), a low-cost 8-bit processor from the American company National Semiconductor. The prototype that he demonstrated to Curry had just 256 bytes of RAM and used a number of components taken from a Sinclair Radionics calculator, including the calculator keyboard and eight-digit LED display.

Curry was suitably impressed by Williamson's demonstration, and Science of Cambridge agreed to develop the product. However, when Curry approached National Semiconductor to negotiate the purchase of the SC/MP in the necessary quantities, the chip manufacturer offered to redesign the device using only National Semiconductor parts. This would considerably reduce production costs but cut Williamson out of the deal.

The result was the Microcomputer Kit 14, a printed circuit board onto which you mounted the calculator keyboard, the necessary chips and the LED display. As with Williamson's prototype, the MK14 came as standard with 256 bytes of RAM, which could be expanded to 640 bytes, and 512 bytes of pre-programmed ROM containing the monitor code. User programs had to be entered in machine code, and there was no permanent storage so your program was lost as soon as you disconnected the power. Perhaps by way of compensation for having come up with the original inspiration, Sinclair paid Williamson £2,000 for the rights to the manual that he had written, which was also supplied with the kit.

The MK14 had more in common with the Altair 8800 than the likes of the Apple II or the TRS-80. It was very fiddly to construct, and, once working, had to be programmed in assembly language. What made it different was the price, with the basic kit costing just £39.95 plus postage. Even with the optional cassette interface and the VDU module, capable of displaying 16 lines of 32 characters each on a standard domestic TV, it came to less than £85. The first manufacturing run of 2,000 proved insufficient to fulfil initial orders, and in total somewhere between 10,000 and 15,000 were sold.

Although determined to build on the success of the MK14, the strategies favoured by Sinclair and Curry (and indeed the NEB) were quite different. While Sinclair felt that the focus should be on producing a usable device at as low a price as possible, Curry felt that reliability was equally important. Curry had become increasingly frustrated with Sinclair's cavalier approach to quality control; a sentiment that Norman Hewett, who had been appointed Managing Director of Sinclair Radionics once the NEB took control, shared:

> Another thing friend Clive isn't keen on is spending money on quality control. Quite often, when there was a meeting where production and engineering needed to be brought together, I wanted to have the quality control manager in as well, because that is of the essence. Clive was always against having him there: "Don't have him in," he'd say. "He just causes problems and trouble!"

> Not only did he not want to be involved in [quality control] himself, but he
> didn't want to know about them or their problems. Extraordinary fellow.
> (Adamson & Kennedy, 1986, p. 56)

Sinclair Radionics had already started work on a microcomputer, but development foundered as the company tried to recover from the Black Watch debacle and Sinclair's loss of interest as he became increasingly involved with Science of Cambridge. The NEB's solution was to focus the company on its core business of scientific instrumentation, selling its calculator and TV interests to Binatone. Clive Sinclair formally resigned from Sinclair Radionics in July 1979, receiving £10,000 in the process, and the NEB passed the microcomputer project to another of the companies that it had funded, Newbury Laboratories, who in 1980 did manage to demonstrate a number of prototypes of what was eventually released as the NewBrain.

Curry was friends with Hermann Hauser, who was studying physics at Cambridge University's Cavendish Laboratory and had helped with the development of the MK14. In 1978, Curry and Hauser joined forces as the Cambridge Processor Unit (CPU), working from a room owned by Sinclair's Science of Cambridge where they developed an improvement on the MK14 based around a 6502 processor – the same as that used by the Apple II and Commodore PET. In March of the following year they launched their new product as the Acorn Microcomputer, available with 512 bytes of RAM, a seven-character LED display, hexadecimal keypad and a cassette interface for £80. Later called the System 1, this was the first product from their new company, Acorn Computers, and was followed by Systems 2, 3, 4 and 5.

Despite the low cost, these were essentially development systems aimed at engineers and scientists. Like the MK14, they came as printed circuit boards and required an understanding of hexadecimal numbers and machine code programming. By contrast the Acorn Atom, launched in March 1980 and largely based on the System 3, looked more like an Apple II or TRS-80, with a proper plastic case and a standard alphanumeric keyboard. It came with a more usable 2K of RAM, which could be expanded to 12K, and an 8K ROM that implemented Atom BASIC, a version of the BASIC programming language that could include statements written in assembly language for speed and flexibility. It could read and write programs to audio cassette, and be plugged into a television set to display 16 lines of 32 characters, monochrome graphics with a resolution up to 256 by 192 pixels, or four-colour graphics at 64 by 192 pixels (although these resolutions were only possible if additional RAM had been installed). It came with a comprehensive manual entitled *Atomic Theory and Practice* and was available in kit form for £120, or for £170 fully assembled.

Sinclair, meanwhile, was still determined to produce something for less than £100, and in January 1980, his company Science of Cambridge launched the ZX80 at a computer fair in Wembley, and also exhibited it at the Consumer Electronics Show in Las Vegas. The ZX80 cost £99.95 fully assembled (equivalent to a little over £300 today), although you did also need a mains adapter, which cost an additional £8.95; or you could buy it as a kit for £79.95. In the US it was advertised as "The first personal computer for under $200" and sold from the company's Boston office for $199.95 or $149.95 in kit form. It was also dubbed "The world's first truly portable computer", and indeed with its distinctive white plastic case it was little bigger than a large paperback and weighed just over 350 grams.

The device came with just 1K of RAM and a 4K ROM that contained a BASIC interpreter specially written to make best use of the small memory space. It could only handle integer arithmetic, but users could also enter machine code instructions. The device had a full alphanumeric keyboard, although this did use a flexible membrane with pressure-sensitive areas rather than proper keys. Like the Atom, it could be plugged directly into an audio cassette for storage and a conventional television to display 24 lines of 32 characters. It could not display colour, but graphical affects could be achieved through special graphic characters that divided each character space into four segments. According to one of its designers, its name was meant to suggest that it added a mystery ingredient, indicated by the 'X', to its Z80 processor.

Sacrifices had been made in order to reduce cost – instead of using a separate chip to process video, for example, this was handled by the main Z80 processor which meant that the display tended to flicker in an annoying fashion – but the result was something that ordinary people could use to enter and run entertaining programs written in BASIC, rather than an arcane machine code. The ZX80 proved an instant success: ten orders were taken within the first five minutes of the Wembley exhibition opening, and the company's office was subsequently deluged with cheques. More than 20,000 units were sold from the UK office over the following eight months (Dale, 1986). Its 1K of memory did severely limit the complexity of the programs it could run, but the option of the ZX80 1-3K Byte RAM Pack (which could be used to add 1, 2 or 3K of memory), followed in September by the 16K RAM Pack, made the device much more attractive. At £49.95, the 16K RAM Pack represented good value, but you did have to put up with 'RAM Pack Wobble' in that the slightest movement could cause the pack to lose power which could mean several hours of laboriously re-typing program code. The usual solution was a blob of chewing gum or Blu-Tack to hold the RAM Pack in place.

March of the following year saw the launch of its successor, the ZX81. This was very similar to the ZX80, using the same processor running at the same clock speed and coming with the same amount of memory. However inside it was very different, as eighteen of the chips used by the ZX80 had been consolidated into a single custom integrated circuit made especially for Sinclair Research (as Science of Cambridge was now known) by the British manufacturer Ferranti. As a result, Sinclair was able to reduce the price to just £69.95 ready-built, or £49.95 in kit form.

The ZX81 proved extremely successful, largely thanks to the deal that Sinclair signed with W.H. Smith allowing the sale of the ZX81 on an exclusive basis through the immensely popular 'computer corners' that the newsagent set up in its high-street stores across the UK. More than 350,000 machines were sold in the first year through W.H. Smith, despite continuing reliability problems. Indeed so many were returned as faulty that the newsagent adopted a policy of routinely ordering a third more than they actually needed (Adamson & Kennedy, 1986, p. 112).

Another reason for the success of the ZX81 was a growing awareness amongst the broader population as to the importance of the microcomputer and the microchip. In 1978 the BBC's long-running documentary series *Horizon* broadcast an episode called 'Now The Chips Are Down', which predicted mass unemployment as a result of the microprocessor, and was followed by a live debate with a government minister discussing the implications. In the following year, computer scientist Christopher Evans published *The Mighty Micro: The Impact of the Micro-Chip Revolution,* which proved so popular that it was turned into a six-part television series broadcast on ITV towards the end of 1979.

In it, Evans suggested that we were now living through the Computer Revolution, and that this would prove to be as significant and traumatic as the

Industrial Revolution of two centuries earlier. The book made a number of predictions as to how computers would transform our lives between 1980 and the year 2000, some of which exhibit the naïve faith in technology that was prevalent at the time, while others have proved eerily prescient. For example, he described with considerable accuracy the modern electronic book reader, but suggested that its widespread use would spell "The Death of the Printed Word" by the early 1990s. He also predicted the end of the "work ethic", with the working week reduced to an average of 30 hours and retirement at 55 or even 50 by the mid-1980s.

Television programmes such as these sparked a debate which resulted in the UK government's Microelectronics Education Programme (MEP) and the BBC Computer Literacy Project. To quote the strategy document written by director Richard Fothergill: "The aim of the [MEP] is to help schools to prepare children for life in a society in which devices and systems based on microelectronics are commonplace and pervasive." The document called for revisions to the curriculum, specialist teacher training and the provision of microcomputers to secondary schools with the intention that "every secondary school will have direct access to at least one microcomputer by the end of 1982" (Fothergill, 1981). The BBC Computer Literacy Project was built around a ten-part television series called *The Computer Programme* and accompanied by a course on BASIC programming plus "a linked microcomputer system complete with User Guide" and a set of specially commissioned application programs (Salkeld, 1981).

The contract to supply the BBC Microcomputer was put out to tender, but the specification was exacting. The project called for something that offered an advanced form of BASIC (at least as good as Microsoft BASIC) while allowing users to enter programmes written in assembly language and to interact with sensors and control electric motors so that they could explore robotics. It also needed to deliver colour graphics and sound.

Research Machines, based in Oxford, was an obvious candidate as the company specifically targeted educational establishments and its 380Z microcomputer was already in use in many schools. Indeed it had already been adopted for the MEP. However the 380Z cost over £500, even in its minimal configuration with just 4K of memory, and the BBC wanted something that the ordinary household could afford. Another candidate, particularly given the amount of public money that had been sunk into Sinclair Radionics over the previous decade, was the NewBrain. This had just been released in prototype form by Newbury Laboratories, after two years in development, and the BBC was keen to adopt it. However it soon became apparent that Newbury was not capable of bringing it into production in time.

Meanwhile Acorn had begun work on a successor to the Atom, provisionally named the Proton. Although this was still at the design stage, Curry managed to persuade the BBC that they could produce a viable machine, and agreed to give a demonstration with just four days' notice. When Hauser first put the idea of completing a prototype in such a short period to designer Roger Wilson, the response was emphatically negative. Hauser then approached Wilson's colleague and friend Steve Furber with the same question, adding that Wilson had already told him he thought it possible. Faced with this challenge, Furber reluctantly agreed, and the two designers worked through the weekend and succeeded in assembling a working prototype, tailored to meet the BBC's requirements, just in time for the meeting. It was a big risk but the ploy worked, and in April 1981, the company was contracted to deliver 12,000 of what was now called the BBC Micro for the Computer Literacy Project (Atack, 1988).

The decision proved controversial. Sinclair in particular was livid, stating in an interview later that year:

> I have no objection to the contract going to Acorn. We have an argument with the BBC on several grounds. First, the way in which it conducted the affair; secondly, selling a product at all; and thirdly, ignoring the industry. (Scott D. , 1981)

However Paul Kriwaczek, producer of *The Computer Programme*, defended the decision:

> At the very beginning, all our advisors told us it was absolutely necessary that an organisation with the standing and public confidence of the BBC should enter the business of computer software, but software for what?
>
> The BBC Microcomputer is expensive, but then a colour television is not cheap either, nor is a stereo system, and we are talking about commensurate kinds of prices. I would have been very reluctant for the BBC to sell something like the Sinclair [ZX81] because it is so limited. The Sinclair cannot be expanded; it is fundamentally a throw-away consumer product. Its usefulness is in learning about programming, but I do not believe that the future of computers lies in everyone learning to program in BASIC. (Gore, 1982)

The BBC Micro used a 6502 processor and came with either 16 or 32K of RAM and a 32K ROM for the BBC BASIC interpreter. It had a proper alphanumeric keyboard and connections for a conventional television, a cassette recorder for storage and a printer. It offered four independent sound channels and eight different display modes, including a 16-colour graphics mode with a resolution of 160 by 256 pixels, and a high-resolution monochrome mode at an unprecedented 640 by 256 pixels (although these modes did require 32K of RAM). The Model A with 16K of RAM was initially sold at £235 while the Model B, with 32K of RAM and facilities for connecting to a wider range of devices, cost £335. The prices were later raised to £299 and £399 respectively.

With the backing of prime-time TV, the success of the BBC Micro was assured. *The Computer Programme* was originally scheduled for broadcast in the autumn of 1981 but was delayed until January 1982 to allow for the production of the hardware. It was followed by *Making the Most of the Micro*, broadcast on BBC2 as ten 25-minute programmes from January 1983, and then *Micro Live*, which ran from 1984 to 1987, all featuring the BBC Micro. Acorn had sold more than 24,000 units by the end of 1982 (twice the number it was contracted to supply), and over a million by the time the BBC Micro was discontinued in 1986. At one point it commanded 60 per cent of the microcomputer market across the UK and Europe.

The success of the BBC Micro was bolstered by an extensive range of supporting software, including many business applications. However games remained important to its success. In 1984 Acornsoft, the software publishing division of Acorn, released a space trading game called *Elite* for the BBC Micro which became immensely popular. The game was written by David Braben and Ian Bell, undergraduates at Cambridge University, at the ages of 19 and 20, and was a masterpiece of programming, pushing the BBC Micro's graphics capabilities to the limit and using advanced techniques to generate over 2,000 planets, each with its own unique set of commodities and prices and all stored within the 32K of memory available on the Model B. Some 150,000 copies were sold for the BBC Micro alone, and the game was eventually converted to many other platforms. Its success even gained it a slot on Channel 4 News when the news editor popped into the newsroom to find nearly all the journalists playing it on their workstations – it wasn't the fact that they were playing computer games, but the fact that they were all playing the same game that caught the editor's attention (Edge Staff, 2009).

The early 1980s saw unprecedented growth in the home computer market. Economies of scale had brought down the cost of memory chips, while advances in technology were increasing capacity. Memory chips capable of storing 16 kilobits were standard and 64 kilobit chips were becoming increasingly popular, particularly after Japanese chip manufacturers slashed unit prices from around $25 to just $5 in 1981 (an 8-bit microcomputer would use eight such chips to give memory capacities of 16K or 64K respectively).

Commodore followed the PET with the VIC-20 in January 1981 (so called because of its purpose-built Video Interface Chip). This became the first computer of any kind to sell over one million units, and was the first with a colour display costing less than $300 (Tomczyk, 1984). The Commodore 64 was launched a year later which, as its name suggests, came with 64K of memory – the most that could be directly addressed by its microprocessor. This was to prove even more popular with sales peaking at some two million a year by the middle of the decade, giving it a 40 per cent share of the market. Like the PET and the BBC Micro, both used the 6502 processor manufactured by MOS Technology, which Commodore owned.

Sinclair's response to the BBC Micro was the ZX Spectrum which, like the ZX80 and ZX81, was based on a derivative of the Zilog Z80 processor that had earlier been employed for the TRS-80. The ZX Spectrum was launched as a mail order product in April 1982 at £125 with 16K of RAM or £175 for the 48K version, making it extremely affordable for a machine with colour graphics and a keyboard which, although based on rubber keys, was a considerable improvement on that of the ZX81. It too proved immensely popular, but Sinclair Research immediately hit supply problems. By August, the company had 40,000 unfulfilled orders, and Sinclair himself apologised publicly for the delay. Unable to properly exploit the Christmas demand, Sinclair changed tack the following year and made the Spectrum available through high street chains such as Boots, Curry's and W.H. Smith, selling 200,000 units by the following March. Then in May, he further stimulated demand by cutting the prices of the 16K model to £100 and the 48K model to £130, and dropping the ZX81 to just £40.

Each computer created its own 'ecosystem' of software and add-ons. You could not take a game written for the Sinclair Spectrum, for example, and expect it to work on a BBC Micro or a Commodore 64. By the summer of 1983, the games market had become fiercely competitive, particularly around the Spectrum and the Commodore machines, where the average title cost just £6. Software companies were appearing and disappearing, selling games with titles like *Atic Atac, Hunchback, Manic Miner* and *Jet Set Willy*, or based around blockbuster films such as *Star Wars* and *Tron*. Distribution companies such as Ocean Software were rejecting 99 per cent of the games they received, many written by teenagers working from home, but would pay an advance of £1,000 to anyone whose game was accepted.

Imagine Software had made a fortune with titles such as *Arcadia*, a big hit of Christmas 1982, and boasted luxury offices in Liverpool complete with a garage stuffed with expensive sports cars. Its founder Mark Butler, just 23 years old, was fascinated by motorbikes, so the company sponsored its own professional racing team. Newspapers were reporting that Eugene Evans, one of the company's young programmers, was being paid £35,000 a year – equivalent to nearly £100,000 today – and had bought himself an expensive sports car, even though he wasn't yet old enough to drive (the story proved to be untrue but reflects the mood of the time). Just 18 months later the company was in financial difficulties, trying but failing to produce a 'megagame' called *Bandersnatch*. It went into liquidation in July 1984 (Andersen, 1984).

Expectations were high as 1984 drew to a close, with sales of £30 million projected for the UK computer games market over the Christmas period. However there were problems. The market was getting swamped with new titles, many heavily promoted but hurriedly and badly designed. Cassettes were easy to copy so software piracy, often on an industrial scale, was rife. By early 1985, newspapers were reporting that the home computer boom was over. In reality it had simply slowed to a more sustainable pace.

Tempers were also running high. In an effort to boost sales of the BBC Micro in the run-up to Christmas, Acorn had placed advertisements in the national press suggesting that the Sinclair Spectrum was unreliable. This was based on market research commissioned by Acorn, but it caused a bit of a ruckus:

> The advertisement so angered Sir Clive that he attacked Curry in the *Baron of Beef*, a Cambridge pub where both are regular customers. Sir Clive walked up to Curry and slapped him about the head, then argued with him about the advertisement. There was some shoving and jostling, and the two men later began fighting again in *Shades*, an upmarket Cambridge wine bar.
>
> Such strong passions amazed the national press, which appear to have believed that the world of technology is populated by cold fish with few emotions. Sir Clive even fell victim to the notorious columnist Jean Rook, who said in the *Daily Express* that she thought the fight gave him sex appeal.
>
> The two leaders of British home computing are now said to have made up

their differences, and Chris Curry was a welcome guest at Sir Clive's New Year's Eve party.

Sir Clive's brother Iain Sinclair comments, "It's nice to know our captains of industry are just as capable of letting their hair down and making complete idiots of themselves as the rest of us." (Sinclair User, 1985)

6.   IBM strikes back

In an ideal world, computer memory would be free, lightning fast and of unlimited capacity. In the real world, at least for the time being, we are saddled with a compromise between capacity, speed and price. The fastest is the silicon chip, and today you can buy a memory stick small enough to fit on your key-ring with a capacity measured in gigabytes (millions of kilobytes) for less than the price of a paperback book. However it wasn't always so. Until only a decade or so ago, memory chips cost serious money and could only really be used for temporary storage within the computer itself. Back in the 1970s, home computers generally used domestic cassette tape for longer-term storage, but these were very slow and unreliable. Mainframe computers used paper tape, punched cards or purpose-built tape systems which were much faster, but prohibitively expensive.

The so-called floppy disk was invented by a team at IBM, initially managed by Alan Shugart, which was tasked with developing a convenient medium for distributing software updates to customers of the System/370 mainframe. The result was a 'memory disk' or 'diskette', eight inches in diameter, coated in a magnetic material and initially capable of storing 80K of data. It was introduced as a commercial product in 1971 but by this time, faced with a promotion which would mean moving from the company's San Jose office to its headquarters in New York some 2,500 miles away, Shugart had left IBM and secured a job at Memorex. (The company's propensity for relocating staff at short notice led some to suggest that 'IBM' stood for 'I've Been Moved'.) Here, he continued to work on floppy disk drives before starting his own company, Shugart Associates, in 1973.

By the mid-1970s, the eight-inch floppy disk drive had become a popular storage mechanism for both mainframe and minicomputers, and indeed MITS did offer it as an option for the Altair 8800 microcomputer, although it was prohibitively expensive at nearly $2,000 in kit form (equivalent to around $8,000 today) or over $2,500 assembled. Shugart Associates was a major manufacturer of floppy disk drives, with its SA800 model establishing something of a standard for the industry. The capacity of an eight-inch disk varied from 250K to over a megabyte, depending on the model, however its main advantage was speed and reliability: a program that might take five or ten minutes to transfer from audio cassette could be loaded from a floppy disk in just a few seconds.

Meanwhile, a number of computer companies had released desktop computers that could be used to perform various office tasks, such as word processing. These were not microcomputers, in that their central processing units were implemented using discrete logic circuits rather than microprocessors, and they were not cheap. IBM had introduced the 5100 Portable Computer in 1975: it had a magnetic tape cartridge storage system and a five-inch monochrome display built into its cabinet. It could be programmed in BASIC and came with cartridges containing three Problem-Solver Libraries covering mathematics, statistics and financial analysis. Prices ranged from $8,975 with 16K of RAM up to $19,975 with 64K of RAM, and there was an optional carrying case to justify the 'portable' tag, although it did weigh some 25kg.

Wang Laboratories, which had already produced a number of word processors, countered a year later with the 2200 PCS (Personal Computer System). This offered similar features to IBM's 5100 but at the rather lower price of $5,400 for a configuration that came with 8K of RAM. The 2200 also used tape cartridges for storage but it did offer floppy disk as an option, and indeed at the time Wang was Shugart Associate's biggest customer.

According to Shugart sales manager Jim Adkisson, one evening he and company president Don Massaro were having a drink in a local bar with Dr. An Wang, founder of Wang Laboratories. Wang was talking about his plans for a lower-cost word processor, and how this would need a floppy disk drive that was not only cheaper but also smaller. Adkisson picked up a cocktail napkin from the bar and said, "Well, what if it was that size?" Wang liked the idea, so Adkisson popped the napkin in his pocket, and the next morning the company set to work developing a prototype. The drive was designed to be the same size as the leading cassette drives of the time, measuring 3.25 inches high, 5.75 inches wide and 8 inches deep, which meant that it could be slotted straight into existing bays. The disk itself measured 5.25 inches square – the same size as the cocktail napkin that Adkisson had pocketed six weeks earlier (Dalziel, Sollman, & Massaro, 2005).

Initially there was considerable resistance within the company because the new product risked undermining Shugart's already successful eight-inch drives. However, as Massaro succinctly put it, "It's better we eat our own children than someone else do it." Accordingly, Shugart launched the SA400 'minifloppy disk drive' in September 1976 at a list price of $390, and Wang came out the following June with the 2200 PCS-II which featured two 5.25-inch floppy disk drives.

As part of its marketing campaign, Shugart showed the new drive to a meeting of the Homebrew Computer Club, where it came to the attention of Steve Jobs. Apple was in the process of launching the Apple II and Mike Markkula, who had recently joined the company as president, saw the provision of floppy disk storage as top priority. However they needed a drive that was cheap, so Jobs paid Shugart a visit. George Sollman, who was product manager at Shugart at the time, recalls:

> After [the Club meeting], the following Wednesday or so, Don came to my office and said, "There's a bum in the lobby, and in marketing, you're in charge of cleaning up the lobby. Would you get the bum out of the lobby." So I went out to the lobby and this guy is sitting there with holes in both knees. He really needed a shower in a bad way but he had the most dark, intense eyes and he said, "I've got this thing we can build ..." (Dalziel, Sollman, & Massaro, 2005, p. 6)

Using a floppy disk for data storage is considerably more complicated than using cassette tape. With cassette tape, all you need to do is read the signal that is coming from the tape head as the tape runs past, convert it into binary data and transfer it into the computer's memory. With a floppy disk, not only is the disk itself spinning at high speed but there is a separate motor moving the magnetic head in and out across a radius so that it can first demarcate and then read and write to the 35 tracks (in the case of the SA400) that run around the surface. Each track is further divided into 10 sectors, each containing 256 bytes of data. The advantage is that any sector on the disk can be accessed very quickly by simply moving the head out to the required track and waiting for the necessary sector to pass by. However this all involves some substantial electronics which, in the case of the SA400, was implemented by 22 integrated circuits.

Steve Wozniak had been studying the manual for the Shugart drive and looking at how other companies such as IBM were using floppy disks. He reckoned that he could come up with something more elegant, so Jobs negotiated for Shugart to supply Apple with a cut-down version of the SA400 that did not include the electronics for just $100 each. Shugart dubbed the device the SA390 to indicate that it was 'less than a standard SA400', and provided Apple with 25 prototypes so that Wozniak could see if his ideas would work in practice.

With typical ingenuity, Wozniak realised that he could harness the computing power of the Apple II itself to do some of the work, and came up with a design that required far fewer integrated circuits. In fact Wozniak had achieved more than he thought, as the drives that Shugart supplied had been rejected for use in the SA400 for quality control reasons (Weyhrich). The result was demonstrated at the Winter CES (Consumer Electronics Show) held in Las Vegas in January 1978, and the Apple Disk II was announced that July at an introductory price of $495, rising to $595 once stock became available. Wozniak later commented, "The disk design was my most incredible experience at Apple and the finest job I ever did" (Williams & Moore, 1985). It also proved to be one of Apple's most profitable products.

As we have seen, floppy disk storage systems give much more immediate access to data than is possible with cassette tape. As a result they lend themselves to more sophisticated storage structures, in that the data stored can be divided into files that can each store different types of information, such as the binary code of a computer program, the text of a document, or the bits and bytes that make up a digital image. Such a structure requires some sort of visual interface so that users can work with their files, allowing them to rename, copy or delete. There also needs to be a programmatic interface so that other software applications, such as word processors or databases, can work with the file system on behalf of their users. The computer program that performs this function, sitting between the user or application that needs to access the files, and the mechanics of the drive itself, is a key component of what is referred to as a disk operating system or DOS.

Gary Kildall had left the University of Washington in 1972 with a doctorate in computer science and joined the United States Naval Postgraduate School in Monterey, California, where he taught computer science. While at the school, he started experimenting with the Intel 4004 processor, which had only just been released, and visited Intel, where he was invited to help develop software for the new device, working as a consultant in his free time. When the 8008 appeared some months later, Intel gave him an Intellec 8 to play with (essentially a microcomputer built around the 8008 processor). This was followed a year later by a sample 8-inch floppy disk drive that Shugart had donated to Intel.

His friend John Torode put together an interface card that would allow the two devices to communicate, while Kildall wrote the code necessary to implement a simple disk operating system. He based the commands on those used by some of the operating systems that came with DEC's PDP-11 family of minicomputers (in particular RT-11) and called his program CP/M, which originally stood for

Control Program/Monitor but quickly came to mean Control Program for Microcomputers.

Kildall offered his work to Intel for $20,000 but was turned down: at that time many at Intel thought that microprocessors would primarily be used to control industrial machines and consumer appliances (indeed Robert Noyce told Kildall that the future for the microprocessor lay "in watches" rather than in personal computers). In response, Kildall and his wife Dorothy formed Intergalactic Digital Research in 1974 with the intention of licensing his operating system to the nascent microcomputer industry. The company name was later shortened to Digital Research (Swaine, 1997).

Kildall had already adapted his CP/M program to work with several different 8080-based microcomputers when in 1976 he was approached by Glenn Ewing of IMS Associates. IMS had sold a number of floppy disk drives to accompany its IMSAI 8080 microcomputer on the basis that an operating system would follow, so Ewing was anxious to strike a deal. Initially, Digital Research sold CP/M to IMS copy by copy, but in 1977 the company granted a licence allowing IMS to distribute the operating system for a one-off fee of $25,000. The agreement established CP/M as the standard for the industry, and other companies soon followed.

The work involved in adapting CP/M to different machines prompted Kildall to separate from his program a component that he called the Basic Input/Output System (BIOS), which contained the code that talked directly to the hardware. This made his life a great deal easier as the BIOS could then be tailored specifically for the machine on which it was to run while the rest of the operating system could remain largely unchanged.

CP/M quickly became attractive to programmers as it made it much easier for them to target a wide range of hardware. If a program needed to access data from a floppy disk, for example, then rather than becoming involved with the intricacies involved in operating a particular disk drive, it could simply send a standard request to the operating system and let that deal with the problem. CP/M also proved popular with users, as once they'd mastered its basic commands they could feel at home on any computer that supported it.

So successful was this approach that, by the end of the 1970s, CP/M was being described as a 'software bus', and the combination of CP/M with the S-100 hardware bus (as originally introduced by the Altair 8800) and an Intel 8080 or

Zilog Z80 processor had become an industry standard supported by almost 100 computer manufacturers, both in the US and the UK. Programs written for CP/M could run on a wide range of microcomputers, which gave a real boost to the software industry. The popular word processing application WordStar was released in 1979 to run under CP/M, as was an early version of the dBASE database manager.

One machine that did not use the CP/M operating system was the Apple II, primarily because it was based around the 6502 processor from MOS Technology, which was not compatible. Instead it ran Apple DOS which had largely been written by Wozniak with help from colleague Randy Wigginton (just 19 years old at the time) and external contractor Paul Laughton. Apple DOS was written specifically for the Apple II and had little lasting effect on the industry. Far more influential was VisiCalc, developed for the Apple II by Dan Bricklin and first shown to the public at the National Computer Conference held in New York City in June 1979.

Bricklin had graduated from MIT in 1973 with a degree in electrical engineering and computer science. He worked as a programmer for DEC for a number of years but found himself more interested in the business world, and so entered Harvard Business School in 1977 with a view to becoming an MBA (Master of Business Administration). As part of the Finance and Production module, students were asked to assume the roles of board members running an imaginary company, taking decisions about employment levels, prices, investment and such like, and working out their effect on turnover and profit. The calculations took a long time using just pencil and paper, so Bricklin started wondering how a computer could be used to speed them up: "Sitting in class, I started imagining the electronic calculator, the word processor that would work with numbers" (Slater, 1987, p. 287). Financial forecasting programs already existed, but these ran on mainframe and minicomputers and were designed to be operated by computer professionals, rather than business executives. What Bricklin was thinking of was something much simpler – something that ordinary people could use.

During the following summer break he made use of the school's timesharing computer to create a prototype of the program's display screen. His starting point was a business calculator with a numeric keypad and a digital display, and he wanted to use some sort of mouse device to operate the program, having seen a demonstration of Doug Engelbart's work some time earlier (see page 102). However, as he worked on the prototype, he realised that a grid of cells arranged into rows and columns would work better, with each cell containing either text, a numeric value or a formula. The formulae could reference values from other cells, which is what gave the program its power.

Bricklin discussed his program with a number of people. The professor of his Production course was encouraging, describing how production planning was often done on tables written "on blackboards that stretch the length of two rooms." Professor Barbara Jackson, who had consulted for the chairmen of several large companies, emphasised that the program would need to be easy to use: "if you want the chairman of the board of a company to do something, it's got to be really simple" (Lammers, 1986, pp. 137-8).

Bricklin also discussed the program with his Finance professor who was not particularly encouraging as he did not feel that the business world would take the microcomputer seriously. However he did suggest that Bricklin get in touch with Dan Fylstra, a former student of his who had gone on to set up Personal Software after he graduated, which he ran from his rented apartment. The meeting went well, and Fylstra lent him an Apple II so that he could develop the idea. Using this, Bricklin wrote a functioning version of his program. It could only work with 100 cells, and supported just the simplest of operations, but it was the first example of an electronic spreadsheet.

Years earlier, while at MIT, Bricklin had agreed with his friend Bob Frankston that the two of them would one day start a business together. About to enter his second year at Harvard, Bricklin called Frankston and suggested they join forces, with Frankston writing a faster and more compact version of the software in machine code while Bricklin worked on the design and documentation. Bricklin and Frankston got to work through the winter producing the new version. Bricklin attended his classes during the day, meeting up with Frankston in the evening to discuss progress. Frankston worked through the night on the code, making use of a commercial timesharing system that charged less after 11pm. The two of them officially formed Software Arts in January 1979, and signed a distribution agreement with Fylstra's company that April. Bricklin had originally called his program Calcu-ledger but this was changed to VisiCalc (as in 'Visible Calculator') following discussions with Fylstra.

VisiCalc didn't make much of an impression at the 1979 National Computer

Conference, or indeed when it was put on sale that October at an initial retail price of $100 (later increased to $150 as sales improved). However it steadily gained influence and it was not long before people were buying an Apple II simply to use VisiCalc. The first known commercial user of VisiCalc was Allen Sneider, a partner at the Boston accounting firm of Laventhol and Horwath. So impressed was he with the program that he persuaded senior management to introduce it throughout the company. The importance of VisiCalc was perhaps best expressed by Ben Rosen, writing at the time in a Morgan Stanley newsletter:

> Though hard to describe in words, VisiCalc comes alive visually. In minutes, people who have never used a computer are writing and using programs. Although you are operating in plain English, the program is being executed in machine language. But as far as you are concerned, the entire procedure is software transparent. You simply write on this so-called electronic blackboard what you would like it to do – and it does it ... We believe that VisiCalc is so powerful, convenient, universal, simple to use, and reasonably priced that it could become one of the largest-selling personal computer programs ever. It's hard for us to imagine any professional user of a personal computer not owning – and frequently using – VisiCalc. (Rosen, 1979)

VisiCalc was initially available only on the Apple II, and can indeed take considerable credit for the machine's popularity. Versions became available for the Commodore PET, TRS-80 and Atari 800 towards the end of 1980, and by 1981 the program was selling 12,000 units a month. By 1983 it had sold over half a million copies (Slater, 1987, p. 291). However its success was to prove short-lived. Software Arts did not immediately produce a version for CP/M, and so were trumped by SuperCalc which Sorcim launched alongside the Osborne I portable computer at the 1981 West Coast Computer Faire. Furthermore relations between Software Arts and its publisher were souring, to the extent that VisiCorp (as Personal Software was now known) sued Software Arts in September 1983 for not meeting contractual obligations.

And so, as the 1970s drew to a close, microcomputers were beginning to make their presence felt in the business world. Busy executives were attracted by the flexibility of VisiCalc, finding it easier to run an ad hoc 'what-if' scenario on their Apple II than going through the formalities of getting the computer department to run it through the company mini or mainframe. As Ted Nelson put it in his usual inimitable fashion:

> Then, like mice, the personal computer arrived. Despite the efforts of the computer center politicians, they came through the windows, over the transoms. And people who bought them in, and could whomp up applications quickly on them, made the computer centers look foolish, which they were. (Nelson, 1987, p. 135)

Those who worked in corporate computer departments were of course familiar with microcomputers – many owned an Apple II or the like on which they would experiment in their spare time. However they were not used to encountering them during office hours. IBM was also aware of the microcomputer, and becoming increasingly concerned at its lack of presence in this new market. The problem for IBM was that it had ruled the industry for decades and was used to setting the pace. IBM was a large corporation, and decisions took time to make and implement. Products had to be carefully developed and thoroughly tested before they were released. As former IBM programmer Rich Seidner put it:

> ... it's like getting 400,000 people to agree what they want to have for lunch. I mean it's just not going to happen: it's going to be lowest common denominator; it's going to be hot dogs and beans. So what are you going to do? IBM had created this process and it absolutely made sure that quality would be preserved throughout the process – that you actually were doing what you set out to do and what you thought the customer wanted. At one point somebody looked at the process to see what it's doing and what's the overhead built into it, found that it would take at least nine months to ship an empty box. (Seidner, 1996)

IBM had no presence in the microcomputer market, which was mutating and expanding with unprecedented speed, driven by tiny companies with track records that could be measured in months. If IBM was to take part, it would need to change its approach.

IBM's General Systems Division had in fact been working on a range of 'low cost' machines that were designed to put more computing power on the desktop for some time. This was the 5100 range, starting in 1975 with the 5100 Portable Computer discussed earlier, and followed by the 5110 Computing System in 1978 and the 5120 Computing System in 1980. They were also working on the System/23 Datamaster, which was to use an Intel 8085 microprocessor (a derivative of the 8080). However, as we have seen, these machines were not cheap: the starting price for an IBM 5120 was $9,340 and a useful system with a printer would set you back well over $13,000. Sales were good, by IBM standards, but they were not competing with the likes of the Apple II. Furthermore they were still subject to IBM's long development cycles: work on the Datamaster had started in February 1978 but it was not actually released until July 1981.

If IBM was to compete in this fast-moving market, it would need to put together something very quickly and very cheaply, both of which ran counter to the IBM culture. One person acutely aware of the problem was William Lowe, laboratory director of IBM's Entry Level Systems (ELS) unit in Boca Raton, Florida, which was also the home of the Datamaster project. He knew that the company would need to take a very different approach from what it was used to, and so he arranged a meeting with the Corporate Management Committee (CMC) in July 1980 at IBM's global headquarters in Armonk, New York State.

He knew he would have to shock the committee into accepting what he had in mind, so his opening gambit was to suggest that IBM's only route into this market was to buy a company like Apple or Atari outright, or buy an existing machine from one of these companies and rebrand it. As expected he was told that such an approach was unacceptable, so he changed tack, suggesting instead that a small task force, operating in a relatively independent fashion and using existing components sourced externally, might be the way forward. Such an approach was not unknown within IBM: after all, the Datamaster was based on a microprocessor supplied by Intel. Furthermore, IBM chairman Frank Cary had introduced Independent Business Units (IBUs) some years earlier, commenting that they "might even teach an elephant how to tap dance" (Chposky & Leonsis, 1988, p. 2). Lowe was given initial approval, and he committed to returning within a month with a working prototype.

Lowe picked 12 engineers for his task force. These included Bill Sydnes and David Bradley, who had worked on the Datamaster project, Jack Sams, who became responsible for sourcing the software, and Lew Eggebrecht who looked after much of the engineering with Sydnes. Once the project was approved, Lowe appointed Philip 'Don' Estridge, who had a long track record with IBM (many years previously, Estridge had worked on some of the software used for the Apollo moon missions) to take charge as senior program manager.

Right from the start it was realised that the nature of the project imposed obvious constraints. The intention was to have the machine ready within a year, so they would need to use off-the-shelf components that were already tried and tested. Furthermore, these components would have to be far cheaper than anything made by IBM if they were to compete with the likes of Apple on price, so the task force would have to source them externally.

This actually tied in nicely with Lowe's decision that the machine should have an open architecture. He was by no means sure that IBM could succeed in this market, but he realised that its chances would be greatly increased if he made it easier for other companies to write software and develop hardware that could work with the new machine, and to do that they would need to know how it worked. This was, after all, the approach adopted for the Apple II, which would be the new machine's main competitor.

Crucial to the project was the choice of microprocessor. Despite the need to cut costs, Lowe wanted a chip that could address more than the 64KB of memory available to processors such as Intel's 8080 or the MOS Technology 6502 found inside the Apple II. Intel had released the 8086 processor two years previously, which used a similar instruction set to the 8008 and the 8080 but with internal and external data buses that were 16 bits wide. Furthermore its external address bus was 20 bits wide, allowing it to address an unprecedented 1MB of memory. In order to achieve this, 16-bit values were extrapolated to 20 bits by combining a 'segment address', which defined the location of a 64KB segment of memory, with an 'offset address' which defined a specific memory location within the segment.

Another possibility was the 68000, released by Motorola in 1979. Internally, much of this processor was actually 32-bit, although its external data bus was only 16 bits wide and internal calculations were only carried out 16 bits at a time. Its address bus was 24 bits wide, allowing it to directly address 16MB without any need for the complications of segment and offset addressing. As such it was far more powerful than the 8086, but totally incompatible and rather more expensive.

Given that members of the task force already had experience of the 8085 processor through their work on the Datamaster project, and given the restrictions on time, Lowe decided to go down the Intel route. However they opted to use the 8088 processor rather than the 8086. Released a year after the 8086, the 8088 was fully compatible but had an 8-bit external data bus rather than the 16-bit bus of the 8086. As a result it could be used with tried and tested (and cheaper) 8-bit memory and support chips, and its 1MB address space seemed more than anyone could need. At the time Motorola was not interested in producing an 8-bit version of the 68000, although it did do so a few years later with the 68008, which found its way into the Sinclair QL.

It was also agreed that the machine would not reach its full potential if it was sold through IBM's usual sales channels, which weren't geared up for the volumes and low margins of the mass market. Instead it needed to appear on the high-street, alongside other microcomputers. With this in mind, Estridge contacted H. L. 'Sparky' Sparks to look after marketing and sales.

So in August 1980, just a month after his original meeting, Lowe returned to the CMC with Sydnes and Eggebrecht, where he presented his plans and demonstrated the prototype (largely built from the parts of a Datamaster housed with a separate keyboard and monitor). The CMC were suitably impressed, upgrading the task force to a full Product Development Group, dubbing the exercise Project Chess and approving the necessary funding for the development of what was provisionally called the Acorn (Chposky & Leonsis, 1988, p. 29). It was agreed that the project should be kept highly confidential – it was vital that the outside world, and particularly the press, did not get wind of the product until it was ready to launch.

Another crucial component was the software, and in particular the operating system. This was Jack Sams' area, and he knew that he should contact Microsoft at the earliest opportunity. By this time Microsoft dominated the microcomputer market as far as computer languages were concerned, and had already created a version of BASIC that ran on the 8086 processor. Microsoft had got together with Tim Paterson of Seattle Computer Products, who had built a prototype 8086 processor board for computers that used the popular S-100 bus, and the two companies had joined forces to demonstrate the result at the National Computer Conference held in New York in June 1979 - the same event at which Dan Bricklin unveiled his VisiCalc spreadsheet program.

Furthermore, Microsoft already had experience of operating systems. The company was very supportive of CP/M as its widespread use made it easier for Microsoft to adapt its computer languages to new machines. Indeed in early 1980, Paul Allen had come up with an ingenious solution to the problem of getting CP/M programs to run on the Apple II by designing an expansion card that contained a Z80 processor. Allen initially talked to Paterson with the idea of getting Seattle Computer Products to produce the board, but Paterson could not get the thing to work, so Microsoft took over development, releasing the result at the West Coast Computer Faire in March 1980. The SoftCard, as it was called, cost $349 and included a copy of CP/M, which Microsoft licensed from Digital Research, together with the CP/M version of Microsoft BASIC. More than 25,000 had been sold by the end of the year.

So late in July 1980, shortly before Lowe demonstrated the prototype of the Acorn to the CMC, Sams contacted Bill Gates to arrange a meeting. Because of the need to maintain secrecy, Sams was not able to tell Gates what the meeting would be about until Gates had signed the necessary Non-Disclosure Agreement (NDA). This was a particularly stringent document stating that if the vendor (i.e. Microsoft) was to disclose anything confidential during the course of their business together, and IBM should act upon the information, then the vendor agreed to not sue. On the other hand, the vendor had to accept that it could be sued if it was to reveal or act upon any confidential information that IBM might choose to reveal. Gates, no doubt intrigued as to why the world's largest computer manufacturer should want to talk to him, signed the agreement. As Sams later recalled, Gates was just 24 years old at the time:

> I knew Bill was young, but I had never seen him before. When someone came out to take us back to his office, I thought the guy who came out was the office boy. It was Bill. Well, I'll tell you or anybody else, and I told IBM executives this the next week, that by the time you were with Bill for fifteen minutes, you no longer thought about how old he was or what he looked like. He had the most brilliant mind that I had ever dealt with. (Wallace & Erickson, 1992, p. 169)

Although it was just a preliminary discussion, Sams was very happy with the meeting, and reported back to the team recommending that they use Microsoft to supply as much of the software as possible. Following Lowe's second meeting with the CMC, Sams arranged a follow-up meeting in which Microsoft was officially told about Project Chess, and commissioned for the job.

Sams knew that CP/M was the dominant operating system in the microcomputer market and, on the evidence of the SoftCard, had assumed that Gates would be able to supply it. Gates informed him that that was not the case, and that IBM would need to negotiate directly with Digital Research. Sams therefore requested that Gates put him in touch with Kildall. Because Kildall had not signed the NDA, Gates was not able to tell him what Sams wanted to discuss, or indeed who Sams represented, but he did suggest that the meeting would be worth Kildall's while. Sams then called Kildall to arrange a meeting for the following Friday. Kildall agreed, but asked that the meeting take place after lunch, as he and his colleague Tom Rolander would need to fly back from a morning meeting that he could not reschedule (Kildall was an enthusiastic pilot and liked to fly his own plane whenever possible).

When Sams and his colleagues arrived, Kildall and Rolander had yet to return and so they were met by Kildall's wife and business partner, Dorothy McEwen, who was understandably ruffled at being presented with a group of corporate executives who wouldn't tell her what they wanted unless she signed a decidedly one-sided legal document. After all, Digital Research had already entered into any number of more conventional confidentiality agreements with its existing customers, so why should these people be any different?

Once Kildall and Rolander did get back, and these initial problems were overcome, further obstacles arose. Kildall wanted to license CP/M to IBM on a royalty basis, receiving a certain amount of money for each copy sold, but Sams proved reticent, preferring to pay a one-off fee for the right to distribute CP/M with the new machine. Sams also wanted to bring the operating system into the IBM brand, which would mean dropping Digital Research's CP/M label. These points particularly concerned Kildall as he was worried that other customers might start pushing for similar terms (Rolander, 2007). In the end, Sams left the meeting doubtful that he would be able to reach an agreement with Kildall. This left him with a serious problem, which he took back to Gates at Microsoft. Gates told him that he would return with a solution by October at the latest.

Meanwhile, Seattle Computer Products had grown tired of waiting for Digital Research to deliver a version of CP/M that it could license for Paterson's 8086-based processor board, so Paterson decided to write his own which he dubbed QDOS (short for Quick and Dirty Operating System). SCP shipped it that September under the name 86-DOS. Microsoft became aware of the existence of QDOS when Paterson contacted Gates to persuade him to produce versions of Microsoft's programs that would run on his new operating system. When Paul Allen heard about the problems that IBM was having with Digital Research, he contacted Rod Brock, owner of SCP, to tell him that he knew of a hardware manufacturer who might be interested in using QDOS, and suggested that Microsoft act as agent. Allen could not reveal who the interested party was because he was bound by the terms of IBM's non-disclosure agreement, and it never occurred to Brock or Paterson that it might be IBM.

So in October 1980, Gates was able to attend the meeting at IBM's Boca Raton plant and announce that he could supply not only versions of Microsoft BASIC, FORTRAN, COBOL and Pascal for Project Chess, but the operating system as well. Gates was happy for the version that would be distributed with the Acorn to be branded as an IBM product, particularly as IBM seemed happy for Microsoft to retain ownership of the source code. As Sams later explained:

> There has been a lot of speculation about why we ever let Microsoft have the proprietorship and all that. The reasons were internal. We had had a terrible problem being sued by people claiming we had stolen their stuff. It could be horribly expensive for us to have our programmers look at code that belonged to someone else because they would then come back and say we stole it and made all this money. We had lost a series of suits on this, and so we didn't want to have a product which was clearly someone else's product worked on by IBM people. We went to Microsoft on the proposition that we wanted this to be their product ... I've always thought it was the right decision. (Wallace & Erickson, 1992, p. 187)

Like Kildall, Gates pushed for a royalty arrangement, rather than a one-off payment that would allow IBM to distribute an unlimited number of copies. This time, possibly because their options were fast running out, IBM agreed. The deal was signed a few weeks later, and Microsoft was given two prototype Acorns on the understanding that it would demonstrate a working version of the operating system and the BASIC interpreter by the middle of January. Work began on adapting 86-DOS to work with the prototype machines which, under IBM's insistence, were kept in a locked room. IBM even sent the occasional inspector to Microsoft's offices to check that security was being maintained.

It was under these conditions that Microsoft negotiated an agreement, signed in early January, that would give Microsoft a non-exclusive right to distribute 86-DOS. Under the terms of the agreement, Seattle Computer Products received an initial payment of $10,000 plus $10,000 for each licence that Microsoft negotiated, or $15,000 if source code was included. SCP did not, of course, know that there would only ever be one such licence, although the agreement did include a clause that stated: "Nothing in this licensing agreement shall require Microsoft to identify its customer to Seattle Computer Products" (Wallace & Erickson, 1992, p. 195). Paterson left SCP in May 1980 to join Microsoft, at which point he was told about the IBM deal; and in July, less than a month before IBM finally lifted the veil, Microsoft bought 86-DOS outright from SCP for just $50,000 (equivalent to about $130,000 today). However the agreement did leave SCP with unlimited rights to use it and access to any improvements that Microsoft might make – concessions which were to have repercussions later.

By this time Digital Research had completed work on a version of CP/M designed for the 8086 processor, called CP/M-86. Understandably furious that Microsoft appeared to be selling IBM an operating system that looked remarkably like his, Kildall instructed his attorney to inform IBM that he intended to prosecute Seattle Computer Products for copying CP/M without his permission. IBM did not want to get involved in a lawsuit at this late stage, so arranged a meeting with Kildall at Digital Research's offices in Monterey. At the meeting IBM capitulated to Kildall's original demands, agreeing to pay a royalty for each copy of CP/M-86 sold and agreeing not to change its name. They further stated that they would not include either operating system with the hardware, allowing the consumer to choose between the two. However this was on the condition that Kildall explicitly agree not to prosecute IBM for copyright infringement in the future. Kildall and his colleague Rolander took a walk down the nearby beach to deliberate, and came to the conclusion that a corporation like IBM was unlikely to succeed in this fast-moving market. On their return, Kildall signed the agreement.

The IBM Personal Computer, or PC, was launched on 12 August 1981, just a year after the project first received approval (its official name was the IBM 5150, but that was quickly forgotten). The starting price was $1,565 which got you the basic unit with 16K of user memory, an interface for a cassette tape player and a monochrome display card capable of displaying 25 lines of 80 characters on a standard TV screen. A more useful configuration with 64K of memory, two 5.25-inch floppy disk drives and a dedicated monochrome monitor cost $3,390. The IBM PC could be bought on the high street through ComputerLand and Sears Business System Centers, and was promoted through a long-running series of adverts featuring a Charlie Chaplin lookalike dubbed 'The Little Tramp'. Director of communications Charles Pankenier explained that they had chosen such a character because "We were dealing with a whole new audience that never thought of IBM as part of their lives" (IBM, 2011).

With regard to software, customers could buy versions of VisiCalc or the EasyWriter word processor from Information Unlimited Software (actually written by John Draper, who had served time for hacking offences under the name Captain Crunch). Peachtree Software had produced three programs for business accounting, and Microsoft had interpreters that would allow you to write programs in BASIC or Pascal, together with a version of the Adventure game. All of these were "carefully selected by IBM" and designed to work with a version of Microsoft's MS-DOS (as 86-DOS was now known) called PC-DOS which was available for just $40. Alternatively, you could buy Digital Research's CP/M-86 for $240. To quote the original press release, "We expect [the availability of CP/M-86] will provide the opportunity for many current applications to be transferred to the IBM Personal Computer with minimal modifications." However, given the similarity between the operating systems, and the far higher price of CP/M-86, it was unlikely to prove popular.

This series of events is often portrayed as the story of how Seattle Computer Products illicitly (or at least immorally) copied Digital Research's operating system and was then tricked into selling it to Microsoft at a cut-down price, while Microsoft and IBM collaborated to deprive Kildall of any revenue or credit. However, the truth is rather less clear cut.

For a start, much of CP/M itself was not original. Indeed there is an article on the Digital Research commemorative website stating that "Gary copied the commands and file-naming conventions from the DEC PDP-10 VMS operating system" (Libes, 1995). In fact, the syntax and vocabulary used by CP/M has rather more in common with the RT-11 operating system, written in 1970 by DEC for the PDP-11. Commands such as DIRECTORY, EDIT, RENAME, SAVE and TYPE work in the same way and can be similarly abbreviated, so both systems will recognise DIR, ED, REN and so forth. Both systems come with the PIP (Peripheral Interchange Program) utility for copying files from one device to another. Both use a similar format for file names with three-letter extensions to indicate file type, so COM indicates an executable command file, BAS a file containing BASIC source code, and TXT a text file.

With regard to QDOS, Paterson himself is quite happy to admit that he copied the syntax and vocabulary of CP/M:

> I just took his printed documentation and did something that did the same thing. That's not by any stretch violating any kind of intellectual property laws. Making the recipe in the book does not violate the copyright on the recipe. I'd be happy to debate this in front of anybody, any judge. To do this did not require ever having CP/M. It only required taking Digital's manual and writing my operating system. And that's exactly what I did. I never looked at Kildall's code,

just his manual. (Wallace & Erickson, 1992, p. 184)

Elsewhere, Paterson states:

> While DRI was free to sue for copyright infringement, the likely success of such action is still controversial at best. The question was whether the published API, used by all the applications that ran under CP/M, was protected from being implemented in another operating system. There are experts who say no, and there are experts who say maybe, but from what I can tell as of 2007 there is yet to be a successful finalized case ...
>
> If tiny Seattle Computer Products had been sued by Digital Research back when DOS was new, I'm sure we would have caved instead of fighting it. I would have changed DOS so the API details were nothing like CP/M, and translation compatibility [making it easier for others to convert their CP/M programs to run on 86-DOS] would have been lost. But in the end that would have made absolutely no difference. No one ever used or cared about translation compatibility. I had been wrong to think it was a valuable feature. (Paterson, 2007)

Furthermore, Microsoft made changes so that, for example, MS-DOS uses the COPY command in place of PIP. It also manages the file system differently, using a software structure called the File Allocation Table (FAT) which Microsoft had originally developed for its BASIC products.

And it is easy to forget that the success of the IBM PC was far from certain. Microsoft was taking a considerable risk in investing the resources it needed to develop MS-DOS and adapt both BASIC and Pascal to run on the new machine within a tight schedule. As Paterson says with regard to the deal struck between Microsoft and Seattle Computer Products:

> We had no idea IBM was going to sell many of these computers. They were a stranger to this business. Somehow, people seem to think we had an inkling it was going to be this big success. I certainly didn't. So buying DOS for $50,000 was a massive gamble on Microsoft's part, a 50/50 chance. (Wallace & Erickson, 1992, p. 204)

The IBM team were also apprehensive about its success. There was a great deal of scepticism within IBM around the whole idea of a 'personal computer', and the team were breaking all the rules by purchasing components from outside the company and not selling the machine through IBM's usual channels. Furthermore, they were working under tight deadlines and great secrecy, so they had to choose their suppliers carefully. It is clear that Sams and his team were impressed by Gates, and that the relationship between the people involved was good. It is also clear that they found it difficult to deal with Kildall, and indeed Rolander has stated that Kildall "didn't like IBM." He continued, "IBM made it possible for corporate America to buy into PCs. I didn't recognise that. I think Bill [Gates] was really astute in recognising that" (Rolander, 2007).

Nevertheless, it was CP/M that first demonstrated the benefits of a common operating system, able to run on a wide range of machines made by many different manufacturers, and establishing a platform that fostered early versions of applications that we still use today. It was Kildall who came up with the idea of separating machine-specific program code into a BIOS, so making such an operating system easier to implement – an approach that is still used now, as you can see from the start-up screen of any modern PC. These days CP/M is largely forgotten, but at the time it played a key role in getting the microcomputer into the office, and would have continued to do so if Kildall had accepted IBM's initial offer.

Over the next few years, Seattle Computer Products did quite well out of its agreement with Microsoft, which allowed it to distribute the latest version of the MS-DOS operating system without paying a royalty. Tim Paterson had also benefited from a similar agreement when he left Microsoft in 1982 to set up his own company, Falcon Technology, building expansion cards for the IBM PC and compatibles. However, by 1985, business was not so good and Paterson was considering selling his company to a group of businessmen who were particular interested in obtaining Paterson's royalty-free MS-DOS licence, which was now extremely valuable. Paterson approached Bill Gates to discuss the matter, but Gates was adamant that the transfer could not take place. After some fraught negotiations, Gates agreed to buy the assets of Falcon Technology from Paterson for $1 million, and Paterson returned to Microsoft as an employee, where he remained almost continuously until 1998.

By this time, Seattle Computer Products itself was in a similar position, so in 1986 Rod Brock also contacted Gates, pointing out in a letter that his agreement with Microsoft was his company's main asset, and continuing, "We believe the value of Seattle Computer to be approximately $20 million. Before making presentations to potential buyers, we want to see if you might have an interest in purchasing Seattle Computer" (Wallace & Erickson, 1992, p. 341). Jon Shirley, president of Microsoft, responded by informing Brock that his licence to MS-DOS was not transferable, so Brock sued Microsoft for the right to sell the licence. The case went to trial in November 1986 and dragged on for several weeks, with Gates himself in attendance much of the time. It was only after the jury had retired to deliberate its verdict that Microsoft finally settled out of court and paid Seattle Computer just under $1 million to buy back its licence to MS-DOS.

7.   Revenge of the clones

The launch of the IBM PC caused quite a stir. Apple took out a full-page advertisement in *The Wall Street Journal* stating "Welcome IBM. Seriously." The advert continued in a comradely tone: "We look forward to responsible competition in the massive effort to distribute this American technology to the world. And we appreciate the magnitude of your commitment." In the first four months after launch, IBM shipped a little over 13,500 units, representing a revenue of around $43 million (Wallace & Erickson, 1992, p. 215). Apple, by contrast, was selling 15,000 to 20,000 units a month, so could afford to be complacent, at least for the time being. However demand was far outstripping IBM's ability to supply, and the IBM PC had been well received by the press, with Gregg Williams' detailed review in the January 1982 edition of *BYTE* magazine concluding, "This microcomputer is as close as I've ever seen to being all things to everybody. IBM should be proud of the people who designed it."

Software began to appear for the new machine, too. Early programs written specifically for the IBM PC included AutoCAD for computer-aided design and Norton Utilities, a set of tools written by Peter Norton for working with floppy disks. The Lattice C compiler brought the powerful C programming language to the IBM PC, and was re-released the following year as Microsoft C under a distribution agreement. Then, in January 1983, Lotus Development Corporation released the spreadsheet program Lotus 1-2-3. Developed by Mitch Kapor and Jonathan Sachs, Lotus 1-2-3 was written in assembly language specifically to run on the 8086 and 8088 processors which made it inherently faster and more powerful than VisiCalc. At $495 it was considered "a tremendous buy for the money", particularly with its ability to generate a variety of graph types on the PC's four-colour display (Williams G. , 1982). The program took over from VisiCalc in giving the business world a reason to buy the IBM PC, and proved so popular that by the end of 1983 – after just one year in business – Lotus was able to report a revenue of $53 million (Power, 2004).

Of course the IBM PC was not the only microcomputer being used in the business world. There were other machines that could run both MS-DOS and

CP/M-86, such as the Victor 9000, designed by Chuck Peddle who was also responsible for the Commodore PET. The Victor 9000 was distributed in Europe by Applied Computer Techniques (ACT) as the Sirius 1, and was followed by the Apricot PC. These machines were not compatible with the IBM PC but were supported by a good range of software, and became particularly popular outside the US where the IBM PC was not officially launched until January 1983. Older 8-bit CP/M machines were also in widespread use, as were the Apple II and the BBC Micro, which now boasted a wide range of business applications and options for adding a floppy disk drive and a Z80 Second Processor. Like Microsoft's SoftCard for the Apple II, these enabled it to run CP/M.

But the IBM PC did boast that important 'IBM' badge, and other manufacturers were starting to realise that they could build something that did exactly the same thing, but faster and for less money. The tight deadlines imposed on its development meant that most of the component parts of the IBM PC were supplied by other companies: the processor came from Intel, the floppy disk drives were made by Tandon Corporation, the power supply by Zenith, the printed circuit board by SCI Systems and the display screen by various manufacturers in the Far East. The keyboard was actually made by IBM and had originally been designed for the Datamaster, but it was easily copied. If your version of the IBM PC was sufficiently compatible, it could run PC-DOS, which could be bought from any IBM dealer. Alternatively, manufacturers could license MS-DOS from Microsoft and then customise it to run on their machines. The only problem was the BIOS: the small but vital computer program that sat between the hardware and the operating system.

For a machine to be fully compatible with the IBM PC, it had to be able to run any software program that could run on the IBM PC, and to work with any expansion card that worked with the IBM PC, and to do that it had to have a BIOS that worked in exactly the same way as that written by IBM for the IBM PC. At first sight this wasn't a problem as IBM had thoughtfully published the program code itself in the *IBM PC Technical Reference* manual, together with the full specification of the machine's five expansion slots. IBM had done this in order to make it easier for companies to write software and develop peripheral devices for its new machine and, certainly, having access to the source code of the BIOS was invaluable, particularly if you were writing in assembly language so that your program would be fast and compact.

However the source code of a computer's BIOS program was clearly protected by copyright: a fact that – by coincidence – had been reinforced only weeks after the launch of the IBM PC. Tandy Corporation had sued Personal Micro Computers on the basis that the "input-output routine" stored in the ROM of the latter's PMC-80 was a copy of that found in Tandy's TRS-80. PMC pushed for the case to be dismissed, arguing that the ROM chip itself was not a 'copy' of the original computer program in a literal sense. However the judge disagreed, upholding the view that a silicon chip was a 'tangible medium of expression' as defined by the 1976 Copyright Act.

So, actually copying the source code was clearly a breach of copyright, although that did not stop less scrupulous companies from doing exactly that. Other companies took a different approach. The functionality of the BIOS program – how it behaved in response to various inputs – was not subject to copyright, so it was perfectly legal to write another program that behaved in the same way (a process known as 'reverse engineering'). The problem lay in being able to demonstrate, possibly in a court of law, that your code was original work and developed with no reference to the work of IBM.

The first company to achieve this was Compaq which had been set up by Rod Canion, Jim Harris and Bill Murto specifically for the purpose of creating a portable computer that would be fully compatible with the IBM PC. The first step in the process was the recruitment of a team of programmers who had never seen the source code for the IBM BIOS, and were prepared to sign affidavits to that effect. The team then created a functional specification for the program, which they determined by subjecting the IBM version to every possible input and noting its response. Once this had been done, they could write a program that implemented the specification, confident that it was original work and in no way breached IBM's copyright. The exercise cost Compaq $1 million and took several months, but the result was the Compaq Portable which became available in January 1983. At 12.5kg and the size of a suitcase it was perhaps better described as 'luggable', but for $3,590 you got 128KB of memory, two 5.25-inch floppy disk drives and a 9-inch monochrome monitor. It became extremely popular in the business world, selling 47,000 units and bringing in $111 million in its first year – at the time a world record for first-year sales achieved by any company (Cringely, 1992, p. 173).

Initially IBM did not seem too concerned about the increasing number of PC compatible or 'clone' machines that were becoming available. After all, it was having a hard enough time fulfilling orders for its own machine, and many of the early clones proved to be far from compatible: some did not contain enough room to take some of the larger expansion cards, for example, or would crash while attempting to run software written for the IBM PC. However by 1983 the situation had become more serious. The production of PC compatible computers had become an industry in its own right, with companies in Taiwan and Japan in particular competing to manufacture compatible parts, such as disk drives and power supplies, and so driving down the price. Many of these Far Eastern initiatives were receiving considerable government backing, too. In 1982 the communist government in Taiwan, for example, funded the Electronics Research and Services Organisation (ERSO) to develop a PC compatible BIOS which was made available to Taiwanese companies free of charge.

IBM began taking legal action against companies that it felt were infringing its copyright. One such was Eagle Computers which was forced to halt production of its Eagle PC until it could develop a legal BIOS, and ended up going out of business. Similar action was taken against Corona Data Systems, which had to redesign the BIOS for its Corona PC, and Handwell Corporation, which was importing machines made in Taiwan that IBM claimed infringed copyright. IBM initially tried to prevent the import into America of any machine that used the ERSO BIOS but in 1984, following negotiations, actually helped ERSO to rewrite the code so it could supply a BIOS that did not breach IBM's copyright (Government Information Office, Republic of China (Taiwan), 1984).

Meanwhile Phoenix Technologies used a process similar to that employed by Compaq, and then persuaded Lloyd's of London to underwrite an insurance policy that would protect anyone that used its BIOS against any litigation that IBM might attempt. The Phoenix BIOS was released in May 1984 as a ROM chip containing the code for just $25. The last barriers to building a respectable and legal PC compatible had been removed. Furthermore, IBM was facing competition from a new breed of compatibles based on the fully 16-bit Intel 8086 processor, rather than the hybrid 8088 used by the IBM PC. The venerable Italian company Olivetti released its M24 in 1983 which proved extremely compatible but considerably faster thanks to its 8MHz clock, in contrast to the slower 4.77MHz clock employed by the IBM PC. Compaq followed suit with a similar machine the following year in the Deskpro. Nevertheless IBM still set the benchmark, and only the IBM PC itself could claim to be truly 100 per cent compatible.

This state of affairs was not unfamiliar to IBM: after all, competitors had long been selling 'plug compatible' peripherals that could be used with IBM mainframes in place of IBM's own products. As long as IBM retained control of the interface – the 'sockets' into which these peripherals plugged – the situation remained acceptable. In much the same way, once the IBM PC proved successful, a primary aim for IBM became the retention of control over the standard by which PC compatibility was judged. For the time being, influential companies like Compaq and Olivetti were looking to IBM to lead the way, but it needed to come up with a convincing strategy.

On the other hand, IBM was a very large company with a broad range of existing products managed by a number of departments, each with its own strategies and architectures, and each subject to the internal politics and personal agendas that inevitably hold sway in such organisations. The IBM PC had been created by a maverick team operating under a very flexible mandate, and many within the company felt that it needed to be properly integrated into corporate strategy. There was also a degree of resentment building up; it didn't help, for example, that the *Wall Street Journal* chose to relegate the heavily promoted release of a new line of mainframe computers to its back pages while devoting prominent up-front coverage to speculation as to what Estridge and his team might get up to next (Chposky & Leonsis, 1988, p. 138).

Meanwhile Estridge and members of the IBM PC development team were drawing up plans for the future. The result was three proposals that were presented to the Corporate Management Committee in February 1982. It was not Estridge's intention that they should all be given an immediate go-ahead as that would spread the company's resources too thinly – particularly as one of the proposals, for what would eventually become the IBM PC AT, was little more than a vague suggestion. Nevertheless, the CMC chose to grant immediate funding for all three.

The least problematic turned out to be the IBM PC XT (also known as the IBM 5160), which was released just over a year later at a price of $4,995. The XT stood for eXtended Technology and its main selling point was the inclusion of a 10MB hard disk. Hard or 'fixed' disk drives had been available for some time, becoming more popular after Seagate Technology (formerly Shugart Associates) introduced the ST-506 in 1980 which was designed to fit into the space reserved for a 5.25-inch floppy disk drive. This was the first time that a microcomputer had been released with one already built in, and it proved popular, but otherwise the

XT was very similar to the original IBM PC.

More problematic was the 'Peanut'. Although supported by quite a few games, the IBM PC was increasingly being used within corporations as a general-purpose workstation. The idea therefore was to create a 'junior' version that would be used in the home and compete directly with the likes of the Apple II. Estridge put Bill Sydnes, who had worked with him on the IBM PC, in charge of bringing the Peanut to fruition. Sydnes approached the project in much the same way as Estridge had approached the IBM PC, and indeed the intention was to operate to a similar schedule, which would mean that the new machine would become available by the summer of 1983 at the latest.

However Sydnes was not Estridge, and was not operating in the same environment. The IBM PC was a shot in the dark, and Estridge had been given an unprecedented degree of autonomy. In contrast, Estridge kept a close eye on Sydnes' progress, and Estridge himself was under scrutiny. Others within the company could see their customers beginning to use IBM PCs to carry out work previously assigned to minicomputers and mainframes, and that was upsetting the delicate support network that underlay IBM's long-standing business model. They were getting worried.

For the 'Peanut' project, Sydnes planned to build something that was fully compatible with the PC and could easily be expanded to offer the full power of the original, but considerably cheaper. He believed that it could be "a dynamite product that could have blown the Apple II series off the map" (Chposky & Leonsis, 1988, p. 144). He also argued that the new machine should be sold through department stores such as Kmart and J.C. Penny, rather than existing channels. Estridge, on the other hand, had misgivings about associating the IBM brand with such outlets, and was uneasy about building a machine compatible enough to steal sales from the IBM PC itself. The project started to run late and encountered increasing problems in design and production. Finally, just as the project was supposed to be reaching completion, Sydnes resigned in exasperation.

While this was happening, IBM was taking steps to integrate the IBM PC more fully into its product line. In August 1983, a new Entry Systems Division (ESD) was created with Estridge as president. This was a substantial promotion, putting Estridge in charge of some 10,000 employees, but it also signalled the coming of age of the IBM PC. Writing in IBM's internal magazine *Currents,* Estridge indicated that he was fully aware of the implicit compromise:

> The business we are generating is large enough to warrant more central
> management in terms of planning and measurement, but we will still be walking
> a line between being an Independent Business Unit and a more conventional
> IBM division.
>
> Personal computing and workstations are very dynamic areas of our industry.
> Advances, enhancements and new products are introduced in rapid fire
> succession. To compete in that marketplace, we have to be able to act and
> respond quickly. Our management style is geared to eliminating overhead,
> unnecessary meetings and discussion; to operating with lower costs, fewer people
> and shorter development cycles. (Chposky & Leonsis, 1988, p. 151)

One of the first tasks facing the new division was to sort out the now leaderless
'Peanut' project. Estridge put Dave O'Connor in charge, and the IBM PCjr (short
for 'junior') was eventually launched in October at a starting price of $699,
although it soon became apparent that machines wouldn't actually reach customers
until after Christmas. The PCjr was fully compatible with its big brother, but did
not have space for more than 128K of memory, which was not enough to run many
important applications such as Lotus 1-2-3, the WordStar word processor and the
popular *Microsoft Flight Simulator*. Furthermore, Estridge had got his way in that it
was being sold through the same retail channels as the IBM PC, but it offered
dealers less of a margin which gave them little incentive to promote it. There was
also the keyboard with its rubber keys which were likened by the press to chunks of
Chiclets chewing gum. As *Time* magazine reported:

> Only six months after IBM introduced the PCjr home computer in
> November 1983, the machine looked like one of the biggest flops in the history
> of computing. Despite IBM's towering prestige and a marketing budget
> estimated at $40 million, the PCjr sold as sluggishly as Edsels in the late 1950s.
> (McCarroll, 1984)

IBM responded in July 1984 by dropping the price, replacing the keyboard
with something more like that of the original PC, and providing facilities for its
memory to be expanded to 512K. Sales did improve, but fell back once the
discounts ended. The PCjr was finally discontinued in March 1985, leaving IBM to
offer its surplus inventory, rumoured to amount to some 100,000 machines, to
employees at a third of the original price.

At this time, most PC compatibles were still using the Intel 8086 processor, despite
the fact that Intel had introduced two successors some years previously. The first,

dubbed the 80186, was very similar to the 8086, although it did integrate many of the external support chips onto the chip itself which made it faster and more efficient. Far more significant was the 80286, released a month or so later in February 1982.

When first powered up the 80286 operated in 'real mode', fully compatible with the 8086 but, like the 80186, considerably more efficient. Real mode used the same technique as the 8086 to address memory, which meant that programs running in real mode could only directly access 1MB. However the 80286 could be switched into 'protected mode' which changed the way it addressed memory, making full use of its 24-bit address bus to directly access as much as 16MB. Furthermore, when in protected mode, each segment of memory and each task that the processor performed was allocated a privilege level. The 80286 supported four privilege levels, and the lower the privilege level the higher the privilege; so a task assigned a privilege level of zero had free rein, but a task operating at privilege level three could only access memory with a privilege level of three, and was prevented by the processor from reading or writing data to memory that had a privilege level of zero, one or two.

Protected mode was designed to tackle a problem that was becoming increasingly irksome. In the early days, microcomputers only had to follow the instructions of a single program. The program might well contain errors or 'bugs' that would cause it to freeze or 'crash', but at least these were the responsibility of a single programmer, or a single software development team. When machines started to use disk operating systems like CP/M or MS-DOS, things got more complicated.

For example, take a look at what happened when you switched on an IBM PC (which is very similar to what happens now when you switch on your desktop or notebook). Computers are hard-wired to start executing their built-in BIOS program as soon as they are powered up. Once the BIOS has checked that everything is present and correct – which it does by executing a series of instructions known as a 'Power-On Self-Test' or POST routine – and has finished a few other housekeeping tasks, it looks around for a suitable 'boot device'. This could be a hard disk, if the machine has one, or a 'bootable' floppy disk that the user has inserted into the floppy disk drive. If it can't find anything suitable, the BIOS displays a message asking the user to insert one. (Incidentally, the term 'boot' is short for 'bootstrap' and refers to the blatantly impossible act of 'pulling yourself up by your bootstraps'.)

Assuming that it finds a floppy disk containing a copy of PC-DOS, the BIOS hands control over to the operating system, which proceeds to load various programs from the disk into pre-assigned locations within the machine's memory. These are the routines that know how to write and read data to the disk drives, how to display data on the screen or send it to a printer, and how to interpret any instructions that you might type in. Once the operating system has been fully loaded into memory, it displays the famous 'A-prompt', indicating that it is ready to execute your instructions.

At this point you can remove the disk containing the operating system and replace it with a disk containing a copy of, for example, the Lotus 1-2-3 spreadsheet application. You then type '123' at the A-prompt, which tells the operating system to look on the disk for a file named '123.EXE', load it into memory and start executing the instructions it contains. This may in turn load other programs from the disk, but in the end it leaves you ready to work on your spreadsheet.

At some point you may decide that you want to save your work to disk, so that you can come back to it later. To do this, you enter the appropriate command into the application which organises your data into the appropriate form, and it then passes control back to the operating system. The operating system in turn passes control back to the appropriate routines in the BIOS, which arrange for your data to be written to your floppy disk.

The problem is that, when running on an 8086 or 8088 processor, all of these programs – the application, the operating system and any other routines that might have been loaded – have full access to all of the memory within your machine. There could easily be a bug in the Lotus 1-2-3 program that, under particular circumstances, causes data to be unintentionally written to an address in memory that happens to belong to the operating system. If that portion of memory contains any of the instructions involved in saving data to floppy disk, for example, then the save operation will go wrong, with the most likely result that the computer will freeze up or 'hang'. The data that you are trying to save – your hard work for the past few hours – may still be sitting in memory, but all you can do to regain control is restart the computer, and your work is lost. As the number of programs sitting in memory increased, and the size and complexity of those programs increased, such mishaps became more common.

However, if you are operating in protected mode, and have an operating system that understands how to use protected mode, then it can operate at a privilege level of zero and assign application programs such as Lotus 1-2-3 a privilege level of three. If this were the case, any attempt by an application to corrupt memory belonging to the operating system would be automatically blocked.

IBM finally launched the IBM PC AT in August 1984, some 18 months behind schedule. The 'AT' stood for Advanced Technology, and in addition to the 80286 processor it came with 256K of memory and a new floppy drive capable of storing 1.2MB of data – four times as much as those that came with the IBM PC. The basic model cost $3,995 (equivalent to around $8,500 today), and there was an option for a 20MB hard disk drive that was twice as fast as the 10MB drive that came with the PC XT. The PC AT also introduced expansion slots that could accept not only the many existing 8-bit cards designed for the original IBM PC, but also a new type of card that could work with 16 bits of data at a time. To do this, IBM simply added a second socket in line with the original PC Bus socket to provide an extra 8 data lines and a further 4 address lines, bringing the total to the 24 lines necessary to access the full 16Mb of memory available to the 80286.

The IBM PC AT was a worthy successor but, like its predecessor, easy to copy. Furthermore, it soon became apparent that there was something wrong with its 20MB hard disk, as users started reporting that it was prone to crashing and losing data for no apparent reason. IBM didn't help matters by denying the existence of the problem, but still halting production. The issue was eventually tracked down to a faulty chip, but the damage to IBM's reputation only served to help the clone manufacturers (Chposky & Leonsis, 1988, p. 175). Compaq had already secured a reputation for quality, and quickly moved in with the Deskpro 286, launched only months later. Other companies followed suit.

IBM had not managed to arrange an operating system for the PC AT in time for the launch that could take advantage of the protected mode offered by its 80286 processor. Instead the machine came with PC DOS 3.0 and a promise that something more powerful would follow. The Entry Systems Division at IBM was in discussion with Bill Gates as to the form that Advanced DOS (as it was being called) might take. The problem was that Intel was close to releasing the 80386, which offered significant advantages over its predecessor, and Gates was keen to work with it. However IBM was adamant that they should concentrate on the 80286.

The 80286 had been designed in 1981 and announced only six months after the IBM PC had been launched, before Intel had any idea of how important the PC would turn out to be, or how many applications would be written for it, all operating in real mode. Intel had expected the industry to follow quite quickly with software that took advantage of protected mode, and so had provided no facility for switching the processor from protected mode back to real mode. Such an option would actually have risked compromising the security offered by protected mode, but this didn't stop Bill Gates from describing the 80286 as "brain dead" (Both, 1997).

What Intel – and everyone else – had failed to foresee was that, by the time IBM launched the PC AT, several million PCs and PC compatibles would be in use, and thousands of real mode applications would have been released. The demand was for a better way to run such applications, and in particular, for the ability to 'multitask'. David Both, who worked at IBM during this period, summed up the situation:

> As soon as I bought my original PC I ran into The Problem. I was writing a letter in *EasyWriter* and needed to make a calculation so I could use the result in the letter. Why should I get out a $10 calculator when I have a $5,000 one sitting here? Of course in order to use it as a calculator, I have to save my document, close *EasyWriter*, reboot to another diskette with the calculator program on it (which I wrote myself in BASIC), do the calculation, write down the answer, reboot to the diskette with *EasyWriter*, load the document, and type in the figure from the paper. (Both, 1997)

This problem could be solved, even on an ordinary PC, by using an MS-DOS function known as Terminate and Stay Resident. Normally, closing a DOS application resulted in control being handed back to the operating system, and the memory that the application had used being marked available for reuse. However an application could also Terminate and Stay Resident (TSR), in which case control would still be returned to the operating system, but the application would remain in memory. Before terminating, such an application could install code that would return execution to itself, perhaps when a particular key combination was pressed. One example was Sidekick, released by Borland in 1984. This provided a pop-up calculator, notepad and calendar, together with a number of other tools, which could be popped up in a panel above another application, such as a word processor or a spreadsheet. Sidekick proved popular, but TSR programs inevitably caused problems, as there was little to prevent one program from corrupting memory that was being used by another.

A more satisfactory solution was the 80386, Intel's first fully 32-bit processor. This contained 275,000 transistors: twice as many as the 80286 and ten times as many as the 8086. It was fully compatible with the 80286 but inherently faster, and in contrast to the 80286, allowed software running in protected mode to directly address an unprecedented 4GB of memory through its 32 address lines. Even more important, the processor could operate in a third mode, called 'Virtual 8086 mode', which allowed it to execute multiple real mode applications side by side. As far as each application was concerned, it was running on an ordinary IBM PC with full access to the machine. In reality it was running in a memory space under the control of the operating system – once an operating system that understood such things became available.

Several reasons have been put forward for why IBM insisted on supporting the 80286, rather than the 80386, at this juncture. By this time the Entry Systems Division (ESD) did not have a particularly good track record, what with the PCjr debacle and the problems with the PC AT that had allowed Compaq to slip in with a perfectly good compatible. There were changes afoot. John Akers took over from John Opel as Chief Executive Officer in February 1985; and in the following month, Don Estridge was promoted to Vice President, which involved moving from Boca Raton to head office in Armonk, New York (he was to die in a plane crash, alongside his wife and several of his colleagues, just four months later). Bill Lowe, who had instigated the original Project Chess, took over as president of the ESD, although he did not move to Boca Raton. Instead he and a selection of other senior members of the division remained near Armonk.

The ESD was being reined in, which meant toeing the company line, and that meant taking note of a set of internal documents known as Corporate Directives. Particularly relevant was Corporate Directive Number 2, signed by Thomas J. Watson Jr. (eldest son of the company's founder) back in 1956 and stating that IBM will keep the promises that it makes to its customers "regardless of the cost." IBM had promised to produce an operating system for the IBM PC AT, with its 80286 processor, and it would do so.

This was very laudable, but the directive had been formulated several decades before the personal computer existed, at a time when the company's customers were major corporations whom they met regularly on a face-to-face basis because each one was investing tens of thousands of dollars every month in IBM products. These new customers were desktop PC users, working in smaller companies, and what they wanted wasn't a corporate strategy but a way to multitask their favourite DOS programs, and build bigger spreadsheets. Such things would be much easier to deliver through an 80386 processor, as Bill Gates well knew.

There may also have been a feeling of unease amongst IBM executives towards the 80386, as it offered the possibility of personal computers that were powerful enough to compete with the company's own mini computers at a fraction of the price. As in every large corporation, the needs of a particular division did not always coincide with what was best for the company as a whole, particularly if the basis of its existence was rapidly becoming redundant.

Microsoft was working on several projects at the time. It had started publishing applications with the Multiplan spreadsheet and the first version of Microsoft Word, and indeed had already demonstrated an early version of Windows (more of which later). MS-DOS was being licensed by an increasing number of clone manufacturers for machines that used both 8086 and 80286 processors, and the company was already looking to adapt it to the 80386. However Microsoft had a close relationship with IBM, and IBM was still regarded as the market leader, defining what it meant to be 'PC compatible'. As Microsoft's vice-president Steve Ballmer later put it:

> It was just part of, as we used to call it, the time riding the bear. You just had to try to stay on the bear's back and the bear would twist and turn and try to buck you and throw you, but darn, we were going to ride the bear because the bear was the biggest, the most important – you just had to be with the bear, otherwise you would be under the bear in the computer industry, and IBM was the bear, and we were going to ride the back of the bear. (Ballmer, 1996)

So, despite the reservations Microsoft had about IBM's approach, in August 1985 the two companies entered into a Joint Development Agreement and started to develop a protected mode operating system specifically for the 80286.

The 80386 processor was announced that October in a version that ran with a 16MHz clock, making it nearly three times faster than the 6MHz 80286 used by the PC AT. Not seeing any reason to wait on IBM, Compaq launched the DeskPro 386 a year later. The machine came with the MS-DOS operating system, so it was essentially just a very fast PC compatible, but at a starting price of $6,499 including 1MB of memory it wasn't a great deal more expensive than a PC AT while, as *PC World* put it at the time, "on CPU performance alone the DeskPro 386 inhabits another league." Other manufacturers followed, and prices dropped. Advanced Logical Research launched the ALR Access 386 for $3,990 which included 512K of memory, or $5,699 with a 42MB hard disk (Satchell, 1986). The mail order company PC's Limited, run by the 21-year-old Michael Dell, released an 80386 machine in March 1987 which *InfoWorld* described as "the fastest desktop computer we've tested" (PC's Limited, 1987).

The market for PC compatibles was expanding in other directions, too. In the UK, Amstrad had already moved into small businesses with a range of CP/M-based word processors, and in April 1986 had purchased the Sinclair brand from Sir Clive Sinclair. That September the company launched the Amstrad PC1512, a PC compatible based on the 8086 processor that cost just £399 including a respectable 512KB of memory, a single floppy disk and a monochrome screen. Even a colour version with a 20MB hard disk cost less than £1,000.

There were other low-cost PC compatibles on the market, but none that could match the specification, and none from such a well-known brand. Furthermore, Amstrad took the opportunity to announce SuperCalc 3, WordStar 1512 and the Reflex database at £69.95 each, fully functional versions of well-known business applications at a fraction of the price. At the same time NewStar Software launched NewWord, described as a 'workalike' to the well-known WordStar 3 Professional but costing £69 instead of £395. Paperback Software announced VP-Planner which worked much like the Lotus 1-2-3 spreadsheet, but for just £99.

The Amstrad PC opened up the world of the IBM PC, with all its supporting software and peripherals, to the small business and even the individual user who might otherwise have bought an Apple II or a BBC Micro. Effectively, Amstrad had succeeded in doing what IBM had tried to do with the PCjr. As Sean O'Reilly of Software Products International put it at the time, the launch of the Amstrad PC marked "the start of a whole new era in the PC world." (Nicholson, 1986).

Meanwhile, IBM was getting bogged down in an attempt to integrate its PC range into its other product ranges through a strategy called Systems Application Architecture (SAA). This grandiose vision defined three services, namely Common User Access, a Common Programming Interface and Common Communication Support, which would be shared across all of its products. The idea was to promote the creation of Common Applications that would share the same user interface and run across any IBM machine, whether it be a top-range mainframe or a desktop microcomputer. IBM wanted the new operating system for the 80286 processor to fully support SAA. However that meant introducing all sorts of complications which were largely irrelevant to Microsoft's main customers, the clone manufacturers.

The result, launched by IBM in April 1987, was a bit of a dog's breakfast. The Personal System/2 or PS/2 range initially comprised four models. The cheapest, at $1,695 (plus $250 for a monochrome display), was the Model 30, which was essentially a well-equipped PC compatible with an 8MHz 8086 processor, 640KB of memory and a 720KB 3.5-inch floppy disk. The Model 50 and 60 both used the 80286 processor running at 10MHz. Only the Model 80 used the 16MHz 80386, and that cost well over $7,000. Furthermore, in an effort to recover the initiative, all but the Model 30 came with a new expansion bus which IBM called Micro Channel Architecture (MCA). This was much more sophisticated than the AT Bus and supported not just 16-bit cards but also, in the case of the Model 80, expansion cards capable of transferring 32 bits of data at a time. However MCA was totally incompatible with what had gone before, which meant that it could not be used with any of the many existing expansion cards available. Furthermore IBM had patented aspects of its design, which meant that clone manufacturers would have to pay for a licence if they wished to use it.

Operating System/2 was announced at the same time at a price of $325, although it did not become available until the end of the year. Developed by Microsoft for IBM under the Joint Development Agreement, OS/2 was designed to work in protected mode on machines based on either the 80286 or the 80386, and supported applications running in up to 16MB of memory. Protected mode applications could be multitasked, which meant you could set more than one running at a time, and switch instantly between them. Such applications could also be multi-threaded so that, for example, you could set a document printing and return immediately to what you were doing without having to wait for the printer to finish, the printing task having been 'spun off' as a separate processing thread. However OS/2 could not take advantage of the virtual 8086 mode offered by the 80386, and so could not run existing DOS applications concurrently. It was also hampered by the need to implement features required by IBM's SAA strategy, which made it cumbersome on any machine that did not boast plenty of memory and a hard disk, and were only of benefit to larger corporations that already had a substantial investment in IBM hardware.

Although some manufacturers did announce that they were going to support MCA, including Olivetti and Michael Dell's PC's Limited, few manufacturers other than IBM actually used it, with most continuing to employ the AT Bus, even for their 80386 models. However, as faster peripherals started to appear, particularly hard disks and graphics cards, the need for a more efficient 32-bit expansion bus increased. Finally, in September 1988, a group of clone manufacturers known as the 'Gang of Nine' and comprising AST Research, Compaq, Epson, Hewlett-Packard, NEC, Olivetti, Tandy, Wyse and Zenith published its own specification for a 32-bit bus called EISA, which stood for Extended Industry Standard Architecture. In contrast to MCA, some 50 companies announced their support for EISA, including Microsoft and the chip manufactures Intel and Phoenix Technologies (the company that had cloned the IBM PC BIOS) (Scannell, 1988).

In practice EISA was only used in high-end machines where performance was paramount. However it had been designed in such a way that an EISA connector could also accept existing 8 or 16-bit cards that had been designed for the IBM PC and PC AT. This was done through the cunning use of slots, so that an ordinary 8 or 16-bit card would bed down only as far as a row of AT-style contacts, while an EISA card could be pushed in further to connect with a lower set of EISA contacts. Of necessity the EISA specification included a specification for the 8 and 16-bit cards that it could accept, and so by implication the AT bus became known as the Industry Standard Architecture (ISA) bus. IBM had lost control of the specification that it had played such a central role in establishing.

IBM continued with the PS/2 range for some time, and even had another go at the home market with a machine dubbed the PS/1. However the steadily increasing power of the personal computer, coupled with the introduction of local area network technology, continued to erode its business. The crunch came in 1993, when in January the company announced that it had lost nearly $5 billion the previous year, followed in July by the announcement of a further unprecedented loss of $8 billion. Judith Dobrzynski, writing in *BusinessWeek*, summed up the situation:

> For all the restructurings, personnel shifts, and product initiatives, the world's biggest computer maker still resembles nothing so much as a flailing giant unable to extricate itself from the mire of an outdated strategy and culture. (Dobrzynski, 1993)

Despite all this, the company did manage to avert disaster by building on its original strengths as a provider of solutions and services to large corporations, and in 2011 was ranked by *Fortune* magazine as the seventh most profitable company in the United States. However it was no longer making personal computers, having sold that part of its business in 2004 to the Chinese manufacturer Lenovo (Williams & Krazit, 2004).

8.   Child's play

All of the computers that we have discussed so far had user interfaces that were little different from those presented by the minicomputers and mainframes that had gone before. Turn on a personal computer that used CP/M or MS-DOS and, once the machine had settled down, you would be presented with a black screen containing little more than a white or perhaps luminous green prompt waiting for you to type in a command and hit the 'return' key. There was no guidance as to what was expected, so you either had to know what you were doing, or have a manual to hand. It was a long way from the ease of use that Alan Kay had envisaged in his 1969 doctoral thesis, and certainly didn't inspire confidence in the ordinary user.

Even once you got an application up and running, there was little consistency between one application and another. The popular word processor WordStar, for example, displayed quite a busy screen with common commands listed at the top; its competitor WordPerfect, on the other hand, opened with an almost completely blank screen. The commands were different, too. Hold down the 'Control' key and tap the 'F' key in WordStar and the cursor would move right one word; do the same in WordPerfect (or press the 'F2' key) and it would think you wanted to 'Find and Replace'. Hold down 'Control' and tap the 'S' key in WordPerfect and it would save your work; to do the same thing in WordStar you had to hold down not only the 'Control' key but also 'K', and then tap the 'S' key. Little consideration had gone into making such applications easy to use, despite the fact that computers were becoming commonplace in the office and the home. Only computer games demonstrated any effort to design user interfaces that were at all attractive or intuitive.

Nevertheless, people had been thinking about how users might interact with computers long before Ed Roberts powered up the first Altair 8800 or Bill Gates typed his first line of BASIC. Much of the early experimentation had been made possible by the unprecedented budgets afforded by the US government to the Advanced Research Projects Agency (ARPA) and the National Aeronautics and Space Administration (NASA) in an effort to win the Cold War. As a result, substantial funding had flowed into universities across the United States through the 1950s and 1960s, in particular into MIT just outside Boston and Stanford Research Institute (SRI) in California, with little consideration as to whether the projects funded had explicit military applications. It was in such projects that some of the visual metaphors that we now take for granted were dreamt up.

An early example was Sketchpad, developed in 1962 by Ivan Sutherland for his PhD dissertation. Working at the MIT Lincoln Laboratory, which had been established by the Air Defense Systems Engineering Committee, Sutherland built a system that allowed people to sketch objects on a display screen using a lightpen, and then have the computer perform operations such as straightening lines or moving one line perpendicular to another. Alan Kay, who studied under Sutherland while he was at the University of Utah towards the end of the 1960s, was strongly influenced by Sketchpad:

> It was not just a tool to draw things. It was a program that obeyed laws that you wanted to be held true. So to draw a square in Sketchpad, you drew a line with a lightpen and said: "Copy-copy-copy, attach-attach-attach. That angle is 90 degrees, these four things are to be equal." Sketchpad would go zap! And you'd have a square. (Rheingold, 1985, p. 235)

Sketchpad was the first application to create the illusion of the computer screen as a window onto a much larger workspace that you could pan across and zoom in to magnify specific details, or out to get a broader view. The program established techniques such as the 'rubber-band line', one end of which remains fixed while the other follows the lightpen around the screen until you click to fix it in place. Sketchpad also allowed you to draw an object and then generate independent copies or 'instances'. In a demonstration filmed in 1962, Sutherland draws a bracket on the screen, followed by a rivet which he inserts into a hole that he has created through one of the flanges of the bracket. He then copies the combined structure a number of times and starts moving each copy independently around the screen, resizing and rotating at will (Sutherland, 1962).

In 1967, Nicholas Negroponte established the Architecture Machine Group ('ArcMac' for short) at MIT with the intention of creating a machine that would facilitate architectural design. Although the machine itself was never completed, ArcMac did manage to demonstrate a 'spatial data management system' which presented the user with a 'dataland' containing representations of data and tools as familiar objects. The user could zoom in to a thumbnail representation of a document, for example, to read its contents. Its most famous manifestation was the 'Aspen Movie Map' which became operational in 1979. Users could take a virtual tour through the city of Aspen in Colorado, viewing images stored on laserdisc and touching on the screen to bring up further information. This actually did have a military application, in that such a system could be used to familiarise soldiers with enemy territory before they were deployed (Cartwright, Peterson, & Gartner, 1999, pp. 20-21).

Another example was the Online System or NLS (the acronym served to distinguish it from the Offline System or FLS). Douglas Engelbart and others started building the NLS in 1962 at his Augmentation Research Center at SRI. The work culminated with a public presentation in December 1968 at the Fall Joint Computer Conference which was attended by Kay and was so innovative that it became known as 'The Mother of All Demos':

> In the darkened Brooks Hall auditorium in San Francisco, all the seats were filled and people lined the walls. On the giant video screen at his back, Engelbart demonstrated a system that seemed like science fiction to a data-processing world reared on punched cards and typewriter terminals. In one stunning ninety-minute session, he showed how it was possible to edit text on a display screen to make hypertext links from one electronic document to another, and to mix text and graphics, and even video and graphics. (Markoff, 2005)

During the demonstration, Engelbart stated that the goal of the project was to "improve the effectiveness with which individuals and organisations work at intellectual tasks." He described the system itself as "a tool for navigating structures that are too complex for direct human study."

The NLS was also the first demonstration of the mouse as a device for selecting items on a computer screen. Engelbart and his team conducted user tests with a number of devices including a joystick, a 'knee control' and a light pen, but found that it was the device we now know as the mouse that offered the best compromise between comfort and accuracy. During the demonstration, Engelbart stated: "I don't know why we call it a mouse. It started that way and we never did change it." The remark was tongue-in-cheek as even the makeshift wooden construction that he was using, with the wire connecting it to the system, brought to mind a dead rodent.

These early experiments eventually led to the graphical user interfaces that we use today. However ARPA and NASA funding was being cut back, mainly thanks to an amendment introduced by Senator Mike Mansfield stating that none of the funds covered by the 1970 Military Authorization Act were to be used for any project or study that did not have a "direct and apparent relationship to a specific military function" (The National Science Board, 2000). Growing opposition to the Vietnam War was also leaving many students unhappy at having research projects within their universities funded by the military.

Meanwhile, spooked by talk of the 'paperless office' that was to come, and in a misguided attempt to compete with IBM, photocopier company Xerox moved into mainframe computers in 1969 by buying Scientific Data Systems for an extraordinary $1 billion. Shortly afterwards, chief scientist Jack Goldman persuaded Xerox to create an advanced research facility, and in January 1970 appointed George Pake as its director. Pake persuaded Goldman to locate the facility in Palo Alto, some 3,000 miles west of company headquarters (which he hoped would prevent too much executive interference), but close to other research centres such as SRI. The Xerox Palo Alto Research Center (PARC) was officially opened that July with a brief to build 'the office of the future', and began to take on researchers. One of the first was Bill English, who had worked closely with Engelbart on the NLS and helped him to orchestrate his famous demonstration. Another was Alan Kay, who was to prove influential in the direction that Xerox PARC would take.

Kay and his colleagues quickly established a campus-style working environment at Xerox PARC, dressing in jeans and t-shirts, bouncing around ideas while lounging around on bean bags, and playing the occasional game of *Spacewar* – an environment that was to cause consternation back at head office when portrayed by Stewart Brand in an article that appeared in the December 1972 issue of *Rolling Stone* magazine under the title 'Spacewar: Fanatic Life and Symbolic Death Among the Computer Bums'. The article did result in stronger security and an insistence that all contact with the popular press be vetted by the press office, but the underlying culture of free-wheeling, open-ended research remained.

At PARC, Kay formed the Learning Research Group whose members included Adele Goldberg and Dan Ingalls. He brought with him his concept of a personal computer which was beginning to consolidate into the Dynabook that he described so engagingly in his 1972 paper *A Personal Computer for Children of All Ages* (Kay, 1972). Although he had been enthralled by Engelbart's 1968 presentation, Kay was thinking of something that would be much easier to use:

> The Engelbart crew were all ace pilots of their NLS system. They had almost instant response - like a very good video game. You could pilot your way through immense fields of information. It was, unfortunately for my purposes, something elegant and elaborate that these experts had learned how to play. It was too complex for my tastes, and I wasn't interested in the whole notion of literacy as a kind of fluency. (Rheingold, 1985, p. 238)

While at the University of Utah, Kay had come across the work of Jean Piaget and Seymour Papert, who had researched the way in which children learn. Piaget had described how children use play to build 'operational models' that they test against the real world and adjust to fit the results. Papert had gone so far as to invent a computer language called LOGO that children could use to program the movements of a small robot called a 'turtle'. In his 1972 paper, Kay argued:

> Piaget's and others' work on the bases and forms of children's thought is a fairly convincing argument for believing that computers are an almost ideal medium for the expression of a child's epistemology. What is an 'operational model' if not an algorithm, a procedure for accomplishing a goal? Algorithms are fairly informal and not necessarily logically consistent ... This fits in well with the child's viewpoint which is global and interested in structure rather than strict implications of 'truths'. On the other hand, the computer also aids in the formation of skills concerning 'thinking': strategies and tactics, planning, observation of causal chains, debugging and refinement, etc. Rarely does a child have a change to practice these skills in an environment that is patient, covert and fun!

At around the same time, Kay had come across a programming language called Simula, which had been created at the Norwegian Computer Centre in Oslo as a tool to aid in the simulation of real-world systems, such as the way in which traffic flowed through a city. Computer programming languages such as BASIC are described as 'procedural languages' because they define a set of procedures that the computer carries out step by step, as you might follow a recipe. Such languages are useful because they work in the same way as the computers on which they run, but become clumsy when you try to simulate a real-world system. Simula, on the other hand, allowed you to create abstract 'classes' that could be programmed to obey certain rules. Using Simula, you could create a class that described how a car might behave, for example, and then generate (or 'instantiate') multiple copies (or 'instances') of your car class and program them to interact so as to model a flow of traffic down a street. For Kay it was a revelation:

... I suddenly realised that Simula was a programming language to do what Sketchpad did. I had never really understood what Sketchpad was. I get shivers now thinking of it. It rotated my point of view through a different dimension and nothing has been the same since. I suddenly understood the purpose of higher-level languages. (Schrage, 1984)

Kay could see how such a language related to the 'operational models' that Piaget had described. The Learning Research Group started thinking about the implications for a machine like the Dynabook. The result was a programming language called Smalltalk which is generally recognised to be the first fully 'object-oriented' computer language, and key to the development of the graphical user interfaces that are so familiar today.

So far the work of the Learning Research Group had been essentially theoretical, as the Dynabook did not as yet exist. However, others within PARC were working on that. In December 1972, Kay's colleague Butler Lampson wrote an internal memo proposing the creation of "a substantial number (10-30) of copies of the personal computer called Alto which has been designed by Chuck Thacker and others." The memo continued, "The original motivation for this machine was provided by Alan Kay, who needs about 15-20 'interim Dynabooks' systems for his education research" (Lampson, 1972). The memo had the desired effect, in that the budget was approved and the first prototypes were ready by the following April.

The resulting machines were quite remarkable for the time. Although not employing a single-chip microprocessor (after all, this was a year before Intel released the 8080 used by MITS in the Altair 8800), the Alto was constructed from integrated circuits. The basic model had a memory capacity of 128K which was eventually expanded to 512K. Other storage was provided by a 2.5MB cartridge system that worked much like a removable hard disk.

The most unusual aspect of the Alto was its display system. The screen measured 8.5 by 11 inches, the same size and shape as standard North American letter format paper, and was orientated vertically. It had a resolution of 606 by 808 pixels which was 'bit-mapped' to 64K of the machine's memory, so that each bit within the 'screen buffer' corresponded to a single pixel in the display.

At the time, documents were usually displayed on a computer screen using 'text mode', a technique that divides the screen into a grid of cells, each containing a single character. A typical screen might contain 24 rows of 80 cells with the character to be displayed in each cell defined by an 8-bit ASCII code. This is very economical, requiring less than 2K of memory to store the contents of the entire screen, but it has limitations. In particular the display font has to be 'mono spaced', in that each character must occupy the same width, and it is impossible to simulate scrolling in a convincing fashion, as characters can only jump from one cell to another.

A bit-mapped display is much more flexible as every pixel can be directly addressed from software. This makes it possible to generate proportional fonts, where characters such as 'i' or 'l' can be narrower than characters like 'm' or 'w'. Characters can be displayed in different sizes, and in bold or italic. Furthermore, text can be moved one pixel at a time, making it possible to simulate smooth scrolling in any direction. The Alto display was good enough to present a reasonable facsimile of ink on paper, but devoting half a machine's memory to such a high resolution bit-mapped display, or indeed spending around $15,000 on building a machine that would only be used by one person at a time, was unprecedented. As Lampson later put it, this was a "… time-machine idea. By spending a lot of money now, you could have the hardware that would be routine and cheap 10 or 15 years in the future." (Lampson, 2006).

As the Alto machines spread through the laboratories, they provided a platform for Kay and other groups within PARC to test their ideas. Bob Metcalfe had been working on a 'local area network' that would be simpler than the ARPANET system that was used to link research centres across long distances (more of which later), and better suited to connecting computers within a single building or across a small campus. The design for the Alto left space for a network controller, and in 1974 Metcalfe was able to fill it with the first Ethernet connectors. Meanwhile Gary Starkweather had completed work on the first laser printer, an extension of Xerox's photocopier technology which he dubbed SLOT (short for 'Scanning Laser Output Terminal'). Pretty soon, SLOTs were being connected to Alto machines and anyone on the network could print to them.

Possibly the most important application for the Alto started life as Bravo, the document editing program developed by Charles Simonyi and Butler Lampson which brought all of this technology together. This was the first WYSIWYG text editor, in the sense that 'What You See' on the screen 'Is What You Get' when you send it to the printer (the acronym was made popular by the Geraldine character created by Flip Wilson for the TV show *Rowan & Martin's Laugh-In* and is pronounced "wizzy-wig"). WYSIWYG was made possible because Starkweather's printer could print to letter-format paper at a resolution that did justice to the

Alto's letter-format screen, with the text displayed on screen visibly matching the font, size and style that it would adopt on paper. Bravo was also revolutionary in that authors could insert text into the middle of a document simply by moving the mouse pointer to the desired location, clicking and typing. However Simonyi had deliberately chosen not to put much work into the user interface in order to save time and resources, so the program was difficult to use.

Which is where Tim Mott and Larry Tesler came in. Mott worked for Ginn, a subsidiary of Xerox which published textbooks, and had been asked to visit PARC with a view to finding a publishing system that could be used by Ginn's book editors. Following discussions with Mott, Bob Taylor (one of the founders of PARC) contacted Bill English to see if he knew anyone who might be able to help Mott develop a suitable system. English put him in contact with Tesler who had joined PARC in 1973 after working at the Stanford Artificial Intelligence Laboratory, where he had helped develop a 'document compiler' called PUB to assist with the formatting and printing of simple documents.

The pair soon realised that Simonyi's Bravo program could do much of what was required, but would need a much more straightforward user interface before the book editors at Ginn would be willing to use it. They started work on what became known as Gypsy by finding out exactly what their intended users needed to do:

> We had just hired a secretary, and the day after she started I put her in front of a blank screen and I said, "Imagine there's a page on the screen, and here's a page of proofreader's marks of what needs to be changed. Imagine that you had a way to point at the screen, and a keyboard. Tell me what you would do." And she said, "Well, I have to delete that. I would point at it, and cross it out. I have to insert some text, so point at the place I want it to go, and I would type it." She just made it up as she went along - what was intuitive to her. (Tesler, 2010)

It is from these early experiments that terms such as 'cut and paste' derive, which are still in use today. However, as they fine-tuned Gypsy, Mott realised that users were finding it difficult to perform basic housekeeping tasks such as filing, printing and emailing that were not part of the application itself. It occurred to him that they could create a representation of an office environment on the screen and let users move documents around this in much the same way as Gypsy allowed users to move text around inside a document:

> I'm sitting in a bar one evening, waiting for a friend to show up, and I'm

doodling on this napkin, and what I came up with was a desk. On the desk was the typewriter, and the editing happened in the in/out trays for handling mail. Separate from that there was an icon that represented a printer, and another one that represented a file cabinet, and one that represented a trash can. And what I saw in this was the ability for people to manipulate objects, almost in a physical way... The majority of the things we control, we control physically - you drive a car, you pick up a pen - and so what we were looking for was an analogue to the way we control a lot of other tools and a lot of other devices in our lives. (Mott, 2010)

The idea was that users could grab the icon representing their document with the mouse and then move it to the icon representing the printer in order to have it print out, or the out-tray in order for it to appear in someone else's in-tray. He labelled the diagram 'Office Schematic' and it became the basis of what users of Microsoft Windows and the Apple Mac came to call 'the desktop'.

By 1976, the network of Alto machines had been distributed to the 50 or so staff at PARC and users were clicking on icons, popping up menus and dragging documents around the screen. Other programs such as Draw and Markup started to appear, building on the work of Sutherland's Sketchpad. An application called Laurel enabled users to compose and send email messages to their colleagues, and was specifically designed to be so easy to use that it did not need a manual. As one user remembers, "At PARC I received my first electronic junk mail, my first electronic job acceptance, and first electronic obituary" (Hiltzik, 2000, p. 212). This was the year in which Apple launched its first computer, and nearly a decade before such facilities would become common in the outside world. Xerox had indeed created the future, but only within the bubble that was Xerox PARC. Thoughts started to turn towards how they might bring all this technology to market.

The starting point was a document drafted by David Liddle, who had been involved in the creation of the Alto, which defined an ambitious 'Office Information Systems Architecture'. Work commenced in 1977, but it soon became apparent that realising the vision was "taking longer than everyone thought and ... was harder than everyone thought" (Hiltzik, 2000, p. 250). Furthermore, many at Xerox headquarters were highly sceptical of the work being done at PARC. After all, Xerox was first and foremost a photocopier company. The company had other problems, too, as the losses incurred as a result of its gamble with Scientific Data

Systems came home to roost and the company was forced to write off $84.4 million.

The Xerox World Conference of November 1977 was an effort to boost morale and herald in a new era through a four-day celebration held at the Boca Raton Country Club in Florida and attended by 250 of the company's executives, together with their wives. The final day of the event was dubbed Future Day and handed over to the people of PARC, who took the opportunity to take a full-blown Alto system, with 30 networked machines, through its paces. Afterwards, the audience was invited to play with the system, but the response was mixed. As one person put it:

> The typical posture and demeanour of the Xerox executives – and all of them were men – was this [arms folded sternly across the chest]. But their wives would immediately walk up to the machines and say, "Could I try that mouse thing?" That's because many of them had been secretaries – users of the equipment. These guys, maybe they punched a button on a copier one time in their lives, but they had someone else do their typing and their filing. So we were trying to sell to people who really had no concept of the work this equipment was actually accomplishing. (Hiltzik, 2000, p. 273)

The product outlined in Liddle's 1977 proposal was eventually launched in April 1981 as the Xerox 8010 Information System, better known as the Xerox Star. No compromises were made: this was a full implementation of the dream that had been demonstrated at Future Day, nearly four years earlier. However, with a single machine starting at $16,595 and a meaningful network costing some $30,000 per user (equivalent to at least $80,000 today), it was simply too expensive. The contemporary executive may have been able to justify the cost of an Apple II and VisiCalc so that he could play around with spreadsheets, but he was not about to spend that sort of money on a system that would primarily be used by his secretary.

And then, just four months later, the IBM PC was unveiled: no mouse, no icons, no windows and no networking as yet, but a fraction of the price and backed by the oldest computer company in the world. In the end Xerox did sell some 25,000 machines, but this was far below initial projections and it is easy to consider the whole Xerox PARC episode a disaster. However the real star was Starkweather's laser printer, which in 1977 was launched as the Xerox 9700 Electronic Printing System and, as Lampson points out, "effectively paid for PARC many times over" (Lampson, 2006).

9. Stealing the TV

Alan Kay's 1972 paper had suggested that the Dynabook would need to cost less than $300 if it was to succeed (equivalent to about $1,500 today). At the time this seemed like an impossible dream, but as the 1970s progressed, many at PARC were aware of the microcomputer revolution that was going on around them. Larry Tesler in particular had attended several Homebrew Computer Club meetings, written a number of articles for the People's Computer Company magazine, and even dated a woman who worked at Apple.

In late 1979, Tesler was invited to a secret meeting with a small group of Xerox executives. As with IBM, many at Xerox realised that the company would never be in a position to compete with the likes of Apple or Commodore because its overheads were too high. As Tesler was told at the time: "If we built a paper clip it would cost three thousand bucks" (Hiltzik, 2000, p. 335). However, rather than take IBM's approach, which would result in the IBM PC, Xerox's strategy was to invest in Apple in the hope that Apple would benefit from the technology and so generate a return on the investment. Specifically, the company had come to an agreement with Steve Jobs which would allow Xerox to purchase a $1m stake in Apple in exchange for them giving Apple technicians a detailed demonstration of the Alto and associated software. Tesler was being asked to organise the demonstration.

Understandably, there was considerable resistance amongst his colleagues to this idea. Initially, Jobs and his team were shown a carefully censored demonstration of the more widely-known aspects of the Alto system, but they soon cottoned on to the fact that much was being withheld and returned a few days later demanding to be shown everything. As Adele Goldberg recalls:

> [Jobs] came back and ... demanded that his entire programming team get a demo of the Smalltalk System, and the then head of the science centre asked me to give the demo because Steve specifically asked for me to give the demo. I said, "no way." I had a big argument with these Xerox executives, telling them that they were about to give away the kitchen sink. I said that I would only do it if I were ordered to do it, because then it would be their responsibility, and that's what they did. (Goldberg, 1996)

The Apple team was spellbound by what they saw. As Jobs later recalled, "... within ten minutes it was obvious to me that all computers would work like this someday" (Jobs, 1996).

This meeting is usually portrayed as the moment when Xerox gave Apple its crown jewels (or Goldberg's "kitchen sink"). However it was a deliberate strategy on Xerox's part, and a strategy that did actually pay off in that, when Apple went public the following year, Xerox's shares became worth over $17 million – although the valuation had rather more to do with the success of the Apple II than with anything inspired by Apple's visit to Xerox PARC.

Apple was on a roll, with sales of the Apple II worth nearly $50 million and growing fast. However this was not an established corporation like IBM or Xerox, with a clear-cut management structure that had withstood the test of time. Instead the company was beset by powerful and eccentric personalities, each pulling in a different direction. As an Apple salesman joked: "What's the difference between Apple and the Boy Scouts? The Boy Scouts have adult supervision" (Isaacson, 2011, p. 155). The situation wasn't helped by Steve Jobs' tendency to polarise relationships, as Bill Atkinson describes:

> It was difficult working under Steve, because there was a great polarity between gods and shitheads. If you were a god, you were up on a pedestal and could do no wrong. Those of us who were considered to be gods, as I was, knew that we were actually mortal and made bad engineering decisions and farted like any person, so we were always afraid that we would get knocked off our pedestal. The ones who were shitheads, who were brilliant engineers working very hard, felt there was no way they could get appreciated and rise above their status. (Isaacson, 2011, p. 119)

As a result, Apple's business strategy was far from coherent. Wozniak felt that there was a lot of life left in the Apple II, while Jobs wanted to move on to something new. Mike Markkula attempted to deliver 'adult supervision' and to this end invited Mike Scott, with whom he had worked at Fairchild, to become president of the company in early 1977. Scott's primary role was to keep Jobs' inspiring but temperamental personality under control.

Perhaps the least coherent of Apple's plans was the Apple III, originally codenamed 'Sara' after the daughter of its chief designer, Wendell Sander. The Apple III was to be aimed at the business market, with the idea that the Apple II could then be specifically targeted at the home and the hobbyist. It used the same

6502 processor as the Apple II, but running at twice the speed, and came as standard with 128K of memory and 140K floppy disk drive. It was essentially a well-equipped Apple II and, although it could run Apple II software, it was not fully compatible as Apple wanted to encourage the development of applications specifically targeted at the new machine.

The Apple III was launched in May 1980 but was beset by problems from the start, and when volume shipments finally got under way the following March, some 20 per cent failed to work because their integrated circuits had dropped out of their sockets during the shipping process. Furthermore, Steve Jobs had insisted that it not have a fan as he felt the noise was too 'industrial', so those that did work frequently overheated. The final blow was the launch of the IBM PC in August, and by the end of 1983, the Apple III had sold just 75,000 units. In contrast, the Apple II had notched up 1.3 million and was still going strong. The Apple III was finally discontinued in April 1984. As one of Apple's founding employees Randy Wigginton later summed up:

> The Apple III was kind of like a baby conceived during a group orgy, and everybody had this bad headache and there's this bastard child, and everyone says, "It's not mine." (Levy, 2004, p. 124)

Thankfully the Apple III wasn't the only new model the company was working on. The Apple Lisa started life as a proposal drawn up by Jobs and Trip Hawkins in October 1978, at around the same time as work commenced on the Apple III. It was named after Jobs' daughter by Chrisann Brennan, although Jobs only confirmed this much later, the official line being that 'Lisa' was an acronym for 'Local Integrated Systems Architecture' (Isaacson, 2011, p. 93). This was to be Apple's first 16-bit computer, intended for general office use and initially planned for release in 1981 at a price of around $2,000. The project was managed by John Couch, who had worked as a software engineer at Hewlett-Packard, and Bill Atkinson was responsible for user interface design.

The Lisa was a far more ambitious project than the Apple III, and even before their visit to Xerox PARC, those working on the Lisa realised that it would need a user interface that would be meaningful to office workers, which meant a bit-mapped display and a processor powerful enough to drive it. There was no obvious 16-bit successor to the 6502 used by the Apple II, and no intention that the Lisa should be capable of running Apple II software, so the initial plan was for Wozniak to design one himself. However in September 1979, Motorola released its first 16-

bit processor in the 68000. Motorola had fallen behind Intel, which had released the 8086 in May of the previous year, so had done much to ensure that the 68000 was the more powerful option. Some of its internal operation was actually 32-bit, and it could directly address up to 16MB of memory. It also allowed programs to run in either User mode or Supervisor mode, a useful mechanism for protecting vital system software, concerned with basic administrative operations, from the vagaries of an application.

Atkinson was aware of the work being done at Xerox PARC and had seen a demonstration of the 'spatial data management system' created by Negroponte's team at MIT, so he had been thinking in terms of a user interface that displayed graphical representations of documents that could be opened into resizable and even overlapping windows. The demonstrations that he, Jobs, Couch and a handful of others from Apple were given at Xerox PARC that December served both to inspire them and to reassure them that such a user interface could be implemented on something like the Lisa. Furthermore the enthusiasm of the Apple team impressed many at Xerox, including Larry Tesler: "After an hour looking at demos they understood our technology and what it meant more than any Xerox executive had understood after years of showing it to them" (Tesler, 1996). Tesler left Xerox the following year to join Apple, working alongside Atkinson on the Lisa, and was eventually followed by some 15 of his colleagues.

Such was his enthusiasm for the Lisa, and in particular its user interface and what he had seen at Xerox, that Jobs started discussing ideas directly with Atkinson and Tesler, without involving Couch. In September 1980, following several disagreements between Couch and Jobs, Scott and Markkula decreed that Jobs be taken off the Lisa project altogether and concentrate instead on being the 'poster boy' for the company's stock market flotation in December – a flotation that was to value the company as a whole at nearly $1.8 billion and Jobs' personal share at $217 million, Markkula's at $203m and Wozniak's at $116 million. Apple was in the money.

Meanwhile in May 1979 Jef Raskin, who had joined Apple the year before, had written an article called 'Computers by the Millions':

> If 'personal computers' are to be truly personal, it will have to be as likely as not that a family, picked at random, will own one. To supply even our own nation with enough computers to make this happen (over, say, a four-year period)

we will require, for example, twenty-five companies each producing over a million computers a year.

The article went on to discuss the problems of manufacturing and supporting computers in those sort of quantities, and the implications for the software industry, coming to the conclusion that computers would need to be built as appliances – sealed boxes that could only be opened by the manufacturer – if companies were to have any chance of managing and supporting such volumes.

Raskin showed the article to Markkula, who requested that he not submit it for publication but instead start work on a machine that might fit the bill. Markkula dubbed the project 'Annie' but Raskin was uncomfortable with the practice of naming computers after women so he changed it to Macintosh, after the apple (actually called McIntosh, but this had already been taken by McIntosh Laboratory, a manufacturer of audio equipment).

Raskin's original proposal was for an 8-bit computer with a minimum of 64K memory based around the 6809 processor from Motorola – the chip on which the 6502, used by the Apple II, had been based. He would have liked it to cost $500 for a minimal configuration, and hoped to have it in production in time for Christmas 1981. An early memo entitled 'Design Considerations for an Anthropophilic Computer' states:

> This is an outline for a computer designed for the Person In The Street (or, to abbreviate: the PITS); one that will be truly pleasant to use, that will require the user to do nothing that will threaten his or her perverse delight in being able to say: "I don't know the first thing about computers," and one which will be profitable to sell, service and provide software for.

Raskin continues:

> If the computer must be opened for any reason other than repair (for which our prospective user must be assumed incompetent) even at the dealer's, then it does not meet our requirements... The computer must be in one lump. This means, given present technology, a 4 or 5 inch CRT (unless a better display comes along in the next year), a keyboard, and disk integrated into one package. It must be portable, under 20 lbs, and have a handle. (Raskin, 1979)

Steve Jobs supported Raskin's vision of a computer as an appliance that could be used by anyone, but could not accept the compromises that Raskin wanted to make in order to keep the price down. At one point Jobs challenged Raskin, "Don't worry about the price, just specify the computer's abilities." Raskin responded with the following memo:

> We want a small, lightweight computer with an excellent, typewriter style keyboard. It is accompanied by a 96 character by 66 line display that has almost no depth, and a ... printer [that] can also produce any graphics the screen can show (with at least 1000 by 1200 points of resolution). In color.
>
> When you buy the computer, you get a free unlimited access to the ARPA net, the various timesharing services, and other informational, computer accessible data bases ...
>
> Let's include speech synthesis and recognition, with a vocabulary of 34,000 words. It can also synthesize music, even simulate Caruso singing with the Mormon tabernacle choir, with variable reverberation.
>
> Conclusion: starting with the abilities desired is nonsense. We must start both with a price goal, and a set of abilities, and keep an eye on today's and the immediate future's technology. These factors must be all juggled simultaneously.
> (Raskin, 1979)

Raskin had fallen foul of Jobs' notorious 'reality distortion field'. This was a trait of Jobs that could be extremely annoying but did sometimes result in the accomplishment of the seemingly impossible. The idea was that if you truly believed something was possible, then it could indeed become possible. As Wozniak put it, "His reality distortion is when he has an illogical vision of the future, such as telling me that I could design the Breakout game in just a few days. You realize that it can't be true, but he somehow makes it true" (Isaacson, 2011, p. 118).

Jobs' contention was that, if the Macintosh was to be "truly pleasant to use" for the "Person In The Street", then it needed a user interface based on what they had seen at Xerox PARC, and that meant a more powerful processor. Without telling Raskin, Jobs asked Burrell Smith, a young engineer working under Raskin on the Macintosh team, to put together a design that used the same Motorola 68000 processor as the Lisa. In fact, Jobs was already in discussions with Motorola.

At the time the 68000 sold for $125 apiece in volume, and a lot more if you only wanted a small quantity. As Thomas Gunter, leader of the team at Motorola that developed the processor, remembers:

> Jobs came in and talked a little bit about the Lisa, and then he said, "But the real future is in this product that I'm personally doing", and that was what became Macintosh. He said, "If you want this business, you got to commit that you'll sell [the 68000] for $15." I said, "Well, what are you willing to commit to?" He said, "Well, we'll use a million a year." I said, "Steve, here's the deal. If you buy a million chips you will not pay more than $15 million, today."

After further negotiation Motorola agreed to stagger the pricing, starting at $55 apiece, dropping to $35 after a certain quantity, and so on. In the end, as Gunter recalls, Apple did buy a million units at an average price of just under $15 each: "In fact, I remember the number pretty well. It was $14.76, not that anybody said" (Brown, Goldman, Gunter, Shahan, & Walker, 2007). Jobs' reality distortion field had worked its magic once again.

By early 1981, the relationship between Raskin and Jobs had deteriorated to the extent that Raskin felt that he had no choice but to raise the issue with Mike Scott. This time, Markkula and Scott supported Jobs, who was given control of the Macintosh project. Raskin took a leave of absence and left the company altogether the following year. Jobs at last had a project that he could really get his teeth into: bringing the technology from Xerox PARC and from the Lisa project to the masses. Something that could truly be described, in his own words, as "insanely great".

Apple's revised business strategy was summarised in a report prepared by Joanna Hoffman at the request of Steve Jobs in July 1981. Called 'Preliminary Macintosh Business Plan', it defined four product lines: the existing Apple II and Apple III models, based on the 8-bit 6502 processor and costing $2,500 and $4,500 respectively; and the Mac and the Lisa, scheduled for release in 1982 and both based on the 68000 processor. The Macintosh would cost around $1,500 for a configuration with 64KB of memory, while the Lisa would start at $5,000 with 256KB of memory. Both would have a bit-mapped display and come with a mouse. The Macintosh would be positioned both as the successor to the Apple II and as "Lisa's younger brother" (Hoffman, 1981).

Jobs rejected the document a number of times, until Hoffman realised that it was not the content that Jobs had a problem with, but rather its appearance. Hoffman had a friend who worked at Xerox PARC and had access to an Alto machine, so she spent a couple of late nights 'borrowing' the Alto to create a document that included representations of drop-down menus and other bit-mapped graphics, which Jobs finally accepted. Particularly ironic was the small graphic representation of a Xerox copier with a line through it and the exhortation 'Do Not COPY' that adorned the lower right-hand corner of the cover.

The Apple Lisa was finally launched in January 1983 with a campaign that boasted, "Apple invents the personal computer. Again." The advertisement continued, "So advanced, you already know how to use it." The Lisa Office System, as the user interface was called, was not dissimilar to that of the Xerox Star, displaying iconic representations of files, folders, a calculator and a wastepaper basket against a grey background. Clicking a file with the mouse opened a window onto its contents, and you could scroll the document within the window, move windows around the screen, and even place one window so it overlapped another, so creating a stack of documents. What was new was the 'drop-down' menu bar running along the top of the screen. Tesler describes moving the mouse pointer along the menu bar:

> As you pass the names of the menus, each menu would pop down one at a time. Going back and forth and they would flutter like that, and the menus would come down, and you go up and down and they would highlight each item. There was just something completely different from anything we'd ever seen before. (Tesler, 2010)

Atkinson continues:

> Having these things flutter along, it feels like they're all there, but they're not really - they're overlapping quite a bit - but it feels like they're all there at any time. (Atkinson, 2010)

At a more fundamental level, the Lisa differed from anything that had gone before in adopting a 'document-centric' rather than 'application-centric' user model. Writing a letter on an application-centric computer such as the IBM PC or Apple II involved loading up a word processing application, creating a new document, saving it using the appropriate command, and then closing the application. To achieve the same result on the Lisa you started by selecting the 'letter' stationary pad from the stationary folder, which opened a window on the screen ready for you to start typing. Once you had finished you could move on to something else, confident that your work would not be lost. Alternatively you could drop down a menu which gave you the option of setting your letter aside, filing it away, or formatting it for print. Behind the scenes the Lisa was continuously saving any changes, and at any time you could simply switch the machine off. Turn it back on and the screen would be restored, ready for you to carry on where you left off. At no point would you consciously have to open or close a program, although in fact you had been working within the LisaWrite application all along. Other applications that came as part of the Office System included LisaCalc, LisaDraw, LisaList, LisaGraph and LisaProject, all developed in-house by Apple.

The Apple Lisa represented the pinnacle of the Xerox PARC vision, and at $9,995 for a basic configuration with 1MB of memory and two 5.25-inch floppy disk drives, was considerably cheaper than the Xerox Star. However it was still far too expensive for its intended audience, many of whom had already invested in the IBM PC. In the event some 13,000 were sold in its first year, which was considered acceptable, but from then on sales failed to meet expectations – particularly after the launch of the Macintosh the following year which provided a cheaper alternative. The Lisa was eventually discontinued in 1986.

The launch of the Apple Macintosh, at Apple's annual shareholder's meeting almost exactly a year after the Lisa, was truly spectacular. Steve Jobs' opening speech ended with the lines:

> It is now 1984 ... IBM wants it all and is aiming its guns at its last obstacle to industry control: Apple. Will Big Blue dominate the entire computer industry? The entire information age? Was George Orwell right?

As he finished, the opening scenes of Ridley Scott's *1984* commercial were playing on the huge screen behind him. Scott, already well-established through *Alien* and *Blade Runner*, had been given a budget of $750,000 to direct a minute-long commercial depicting an Orwellian dystopia ruled over by a Big Brother who is speaking to his subjugated audience. All is grey and dull until a tanned athletic young woman dressed in a white top and bright red shorts runs down the aisle and flings a huge hammer at the screen, shattering the face of the leader into an explosion of light. As the film closes the text appears: "On January 24th, Apple Computer will introduce Macintosh. And you'll see why 1984 won't be like '1984'." The film had already been broadcast as part of the televised coverage of Super Bowl XVIII two days earlier. The slot had cost $800,000 but reached an audience of some 40 million.

As the film finished, Jobs returned to the stage to introduce the machine itself, which he lifted out of its carrying case and placed on a plinth centre stage. His introduction concluded:

> We believe that Lisa Technology represents the future direction of all personal computers. Macintosh makes this technology available for the first time to a broad audience – at a price and size unavailable from any other manufacturer. By virtue of the large amount of software written for them, the Apple II and the IBM PC became the personal-computer industry's first two standards. We expect Macintosh to become the third industry standard. (Regis McKenna Public Relations, 1984)

Finally, with the theme to *Chariots of Fire* playing in the background, Jobs said, "But today, for the first time ever, I'd like to let Macintosh speak for itself." Using a program called SAM the Software Automatic Mouth, an electronic voice responded:

> Hello, I am Macintosh. It sure is great to get out of that bag! Unaccustomed as I am to public speaking, I'd like to share with you a maxim I thought of the first time I met an IBM mainframe: never trust a computer that you can't lift! Obviously, I can talk, but right now I'd like to sit back and listen. So it is with considerable pride that I introduce a man who has been like a father to me ... Steve Jobs!

The audience went wild, with Jobs (and the Macintosh) receiving a standing ovation that lasted several minutes (Hertzfeld, 2005, p. 217). Other efforts to promote the Macintosh included an 18-page brochure inserted into the December 1983 editions of high-circulation magazines such as *Time* and *Newsweek*, and the purchase by Apple of all 39 pages of advertising in a special election issue of *Newsweek* that appeared in November 1984. The opening spread showed the Macintosh from the front, while the back cover showed a view of the rear.

In all but price, the Macintosh remained true to Raskin's original vision, right down to the handle at the top. Although not fully portable, in that it did not include a battery, it did offer an optional carrying case and, as the glossy, colourful manual stated, "Your Macintosh will fit under the seat in most commercial airlines and in the overhead compartment of others." As Raskin had dreamed, and as the official press release stated, "Macintosh was designed from the start to be built in the millions." To that end it was manufactured in a highly automated factory that Apple had built in Fremont, California with an emphasis on quality control, and was sold as a sealed box with "no user-serviceable parts". It was also unusual in that it had been specifically designed for an international market, so icons were used in preference to text to depict the function of keys, controls and ports, and the software was designed so that it could easily be configured to use a different language.

Unlike many in the industry, Jobs had always understood the importance of good design, and was particularly enamoured with the straightforward, functional style of the Bauhaus movement. He had given a talk at the 1983 International Design Conference in Aspen entitled 'The Future Isn't What It Used to Be', in which he had talked about the company's approach:

> We will make [our products] bright and pure and honest about being high-tech, rather than a heavy industrial look of black, black, black, black, like Sony ... Very simple, and we're really shooting for Museum of Modern Art quality. The way we're running the company, the product design, the advertising, it all comes down to this: Let's make it simple. (Isaacson, 2011, p. 126)

Jobs was building up to what would become his signature achievement, and the key to Apple's long-term success, namely the creation of a unique and complete experience that would dictate every interaction between the customer and the brand. He agonised over seemingly trivial aspects of the design, such as the size of the bevels and the colour of the case. Many years previously, he had attended a calligraphy class which brought home to him the importance of the typefaces that would be used by the Lisa and the Macintosh, and so great care was taken to design the Chicago, New York and Geneva fonts that would be displayed on screen. He insisted that the Macintosh be delivered in a full-colour box, and spent an inordinate amount of time and money on its design.

The Macintosh did benefit from the same Motorola 68000 processor as was used in the Lisa, but running rather faster with a clock speed of just under 8MHz as against the 5MHz used by the Lisa. It also came with the new 3.5-inch Micro Floppy Disk drive that Sony had developed, which used smaller disks encased in a rigid plastic shell that could be tucked into a shirt pocket. Like the Lisa, it had a bit-mapped display, but smaller at nine inches across and a resolution of 512 by 342 pixels.

Although Jobs had lambasted Raskin for his small-minded approach, he did accept that compromises had to be made if the Macintosh was to remain viable at a price that the 'person in the street' could afford. The biggest was memory capacity, made necessary by the high price of memory chips. The Macintosh Operating System and the user interface, adapted from the Lisa, was slimmed down into 64KB of ROM within the machine, so that once the machine was switched on and booted up, you were presented with a screen inviting you to insert a floppy disk. However it only came with 128KB of memory and there were no facilities for expansion, although some people did replace the sixteen 64K-bit memory chips with more expensive 256K-bit chips as they became available. This gave the machine a more usable 512KB of memory, but it was an expensive and somewhat risky operation. There was a socket for an optional second external floppy drive, but as yet no option to add a hard disk.

The Macintosh exhibited a user interface similar to that of the Lisa – mouse-driven with icons, windows and drop-down menus – but its limited memory capacity meant that it could only run one major application at a time, which made it impossible to implement the Lisa's 'document-centric' user model. Writing a document on the Macintosh meant loading the MacWrite application from the appropriate 3.5-inch floppy disk, opening a new file, writing the text and remembering to save it to another floppy disk, much as you would with an IBM PC or an Apple II. Saving a document had to be an explicit command, as it needed the user to check that the correct disk was in the drive. That said, it did share with the Lisa an area of memory called the 'clipboard' into which data could be copied and retained ready for access by another application. There were also a number of simple 'desk accessories' that could be run alongside a main application like MacWrite. These included Calculator and Clock, together with Scrapbook which gave you access to the clipboard.

Reviews were generally good, with *BYTE* concluding, "The Macintosh brings us one step closer to the ideal of computer as appliance" (Williams G. , 1984). Towards the end of his detailed review in *Creative Computing*, John Anderson stated, "It should be obvious to you now that the Mac does represent a significant breakthrough, both in hardware and in software." However he went on to list a considerable number of reservations, including a lack of memory, concluding, "The bottom line on this point is that it might be two years or so before you can inexpensively give your Mac enough RAM to be truly useful" (Anderson J. J., 1984). The real stumbling block – and the point at which the Macintosh deviated the most from Raskin's original vision – was price. It was indeed destined to become an industry standard, but for the moment, at $2,495 for a single-floppy machine with just 128KB of memory, little software and few options for expansion, it was just too expensive.

For Microsoft, the first half of the 1980s proved challenging. The company's main revenue came from licensing MS-DOS and Microsoft BASIC to the various manufacturers of PC compatibles. However its principal customer, IBM itself, was starting to explore strategies that would stifle the clone manufacturers, and was expecting Microsoft to help. If IBM succeeded, the most likely outcome would be that clone manufacturers would try and establish their own platform, which risked leaving Microsoft out in the cold. However Microsoft was loath to abandon its relationship with IBM, which had so far proved well worth the considerable resources that Microsoft had put into it.

It was also becoming apparent that the success of a platform depended largely on the applications that support it – after all, most of Microsoft's customers wanted their computers to help them with their day-to-day business, not play around with operating systems or dabble in programming. MicroPro was enjoying considerable success with its WordStar word processing application, with a version for MS-DOS coming out in 1982. A competitor was WordPerfect, which also came out in an MS-DOS version in 1982. There was the dBASE database manager from Ashton-Tate, but the most significant was VisiCorp's spreadsheet application, VisiCalc. Having enjoyed unparalleled success on the Apple II, VisiCalc had been rewritten to run under MS-DOS and was actually released alongside the IBM PC.

If Microsoft did not get involved in application development soon, it risked losing control of its main revenue stream, namely MS-DOS itself. The breakthrough came in February 1981 when Charles Simonyi, who like so many at Xerox had become disillusioned with the company's inability to exploit its own innovations, left to join Microsoft, where he founded its Application Division. He was joined in May by his colleague Richard Brodie (Brodie, 2000). Simonyi described his first meeting with Gates:

> It was amazing. Bill was like twenty-two, looking seventeen. I was thirty-two. The bandwidth we had and the energy just flowing from him was incredible. In a five-minute conversation we could see twenty years into the future ... We are talking about a sunset industry and a sunrise industry. It was like going into the graveyard or retirement home before going into the maternity ward ... You could see that Microsoft could do things one hundred times faster. (Hiltzik, 2000, p. 359)

In the process, Simonyi invited Gates down to Xerox PARC and showed him the Alto in action, printing him a document that demonstrated all the Alto's main abilities while charting a strategy for Microsoft to enter the application market. Gates in turn introduced Simonyi to the IBM PC, which at the time was still a carefully guarded secret.

The division's first application was Microsoft Multiplan, a competitor to VisiCalc that was released in February 1982 for CP/M and the Apple II, and a year later for MS-DOS. Multiplan was followed by Multi-Tool Word, which in many ways was a successor to the original Bravo program that Simonyi had developed at Xerox PARC. It made use of the PC's 640 by 200 pixel monochrome display mode to achieve a WYSIWYG display – the first word processor to do so on a microcomputer – and could be controlled using the Microsoft Mouse, which was released at the same time. The program was also unusual in that you could open up to eight windows, each displaying a different document, and move blocks of text between files using the mouse. Both the program and the mouse became available in May 1983.

Microsoft already had a relationship with Apple through Applesoft BASIC, which Microsoft had written for the Apple II in 1977 to replace Wozniak's original Integer BASIC. Microsoft had also produced the SoftCard, which allowed CP/M programs to run on the Apple II. However Microsoft was a much smaller company than Apple at this stage, with sales for 1981 amounting to $15 million, against Apple's $334 million (Poole, 2002). Apple was aware of Microsoft's intention to develop Multiplan and Multi-Tool Word, and knew that the secret to the Macintosh's success would lie in the applications that it could run; so in July 1981, Steve Jobs gave Gates a demonstration of a prototype of the Macintosh (ironically just before the launch of the IBM PC, which presumably prevented Gates from mentioning its imminent existence). In January of the following year, Gates signed an agreement to produce versions of both applications for the Macintosh, and received prototypes of the machine under conditions of strict secrecy. They were nicknamed SAND, which stood for 'Steve's Amazing New Device' – Jobs having apparently boasted that he would build factories that would turn sand (rich in silicon) into Macintoshes (Carlton, 1997, p. 28). At around the same time, Microsoft started work on a project codenamed 'Interface Manager'.

Although he was already familiar with the concepts behind a graphic user interface thanks to his relationship with Simonyi, signing the agreement brought home to Gates the importance of a consistent platform to the success of Microsoft's nascent Application Division. Yes, MS-DOS was a platform of sorts, but it did not offer the rich user experience of the Xerox Alto or the Apple Macintosh. Interface Manager would work as an extension to the MS-DOS operating system, allowing users to launch programs such as Multiplan and Multi-Tool Word in an intuitive fashion and control them using the Microsoft Mouse. It could also build on Multi-Tool Word's ability to cut and paste between open documents by allowing users to do the same between documents that had been created by different applications.

Then, in November 1982, Gates attended the Comdex trade show in Las Vegas, where he came face-to-face with Visi On. Unknown to Microsoft, VisiCorp had been planning the same thing but was rather further down the road – already at the stage where it could demonstrate a friendly on-screen environment from which users could launch not just the already successful VisiCalc but any other programs that VisiCorp or potential partners might write. Gates was in a difficult position as Interface Manager was at a very early stage, and he had agreed with Jobs not to release any kind of graphical user interface until at least a year after January 1983, which was when the Macintosh was scheduled to ship (Isaacson, 2011, p. 177). However it soon became apparent that the Macintosh was running late, so when January passed with no sign of the machine, Gates felt free to drop hints to the press that Microsoft was working on something that would compete with Visi On.

As the year progressed, work proceeded on Interface Manager, with Microsoft recruiting three further people from Xerox PARC, including Simonyi's friend Scott MacGregor who had worked on the user interface of the Xerox Star. News came that other companies were thinking along similar lines, including IBM with TopView and Quarterdeck with DESQ, while the June issue of *Byte* magazine featured a detailed interview with VisiCorp's William Coleman, clearly in preparation for the imminent release of Visi On (Lemmons, 1983). Even though Interface Manager was far from ready, the pressure was on Gates to make a formal announcement soon.

One consideration was the product's name. Most at Microsoft were happy to call it Interface Manager but Rowland Hanson, who had recently been appointed to handle Microsoft branding and marketing, was not convinced:

> I told my group to go out and gather all the editorial that's been written about all these other GUIs that are out there: "We're going to look at all this stuff, and we're going to look for commonality in how these GUIs are being talked about." So we taped everything to the wall – hundreds of articles around the room – and we started to find a very consistent theme between all these GUIs: they were not being described as 'graphical user interfaces'; they were being described as 'windowing systems'. And I said, "If these are all windowing systems, there's only one name we can pick: we have got to be Windows." It was a Eureka moment but we thought about the name for another six months because nobody wanted to call it Windows ... But it was absolutely the right name at the right time ... (Call MD Plus, 2012)

Then, at the end of October, VisiCorp released Visi On. Gates felt he could wait no longer and so, on 10 November 1983, formally announced Microsoft Windows at a lavish press conference in New York. The demonstration was pretty flimsy, but IBM's decision to put together its own windowing system had actually worked in Microsoft's favour, as it helped Gates to gain the support of PC compatible manufacturers worried that IBM would attempt to set a standard around TopView. Some 24 computer manufacturers who were already licensed to use MS-DOS, including Compaq, Digital, Hewlett-Packard, Texas Instruments, Wang and Zenith, pledged their support for Microsoft's forthcoming Windows.

What was demonstrated was a windowing system based on block graphics and text, lacking the bit-mapped graphics and proportional fonts exhibited by the Xerox Star or Apple Lisa. The windows themselves did not overlap but were rather 'tiled' to sit next to each other on the screen. However, as Microsoft's Leo Nikora stated, "The window manager is actually a small part of this announcement. Windows is actually an attempt to widen the development environment for MS-DOS so that applications can be graphically oriented." What Windows would offer was "complete device independence and a graphical interface for MS-DOS" (Markoff, 1983). This would be achieved through a Graphics Device Interface (GDI) that would allow applications to take full advantage of devices such as high-resolution printers.

DOS applications that were 'well behaved', in that they used the operating system to handle the file system and the display, would be able to run within tiled windows and take advantage of all that Windows had to offer. Where an application was not so well behaved, as was the case with Lotus 1-2-3 or Microsoft's own *Flight Simulator,* which interacted directly with the display hardware, Windows would hand control over to the application when it was launched, and only resume control after the application was closed. This was in contrast to Visi On, which could only run software that had been specifically written for the environment. Another advantage was that developers could use an ordinary PC to create applications for Windows. Visi On applications had to be developed using an expensive development environment that ran on a DEC VAX minicomputer (Markoff, 1983). Microsoft stated that it expected to ship Windows the following April, and Gates boldly predicted that it would be in use on nine out of ten PC compatibles by the end of 1984 (Wallace & Erickson, 1992, p. 258).

When Jobs heard about the Windows launch, just a few months before the Macintosh was to be announced, he was furious and summoned Gates to Apple headquarters. As Andy Hertzfeld recalled, Gates stood alone, surrounded by Apple employees:

> I was just a fascinated observer as Steve started yelling at Bill, asking him why he violated their agreement: "You're ripping us off!", Steve shouted, raising his voice even higher. "I trusted you, and now you're stealing from us!" But Bill Gates just stood there coolly, looking Steve directly in the eye, before starting to speak in his squeaky voice: "Well, Steve, I think there's more than one way of looking at it. I think it's more like we both had this rich neighbor named Xerox and I broke into his house to steal the TV set and found out that you had already stolen it." (Hertzfeld, 2005, p. 191)

Gates certainly had a point. His exposure to the Macintosh at an early date may have inspired him, but the technology behind Windows came primarily from Xerox employees who were now on his staff. Moreover, Microsoft did release Multiplan for the Macintosh shortly after the machine launched, and Microsoft Word the following year. Nevertheless, Gates had deliberately misled the industry. Windows was far from ready and would not actually be released until November 1985, a full two years later. At the time of its New York debut, it was a prime example of what was increasingly becoming known as 'vapourware'.

10.   The killer app

Mike Markkula replaced Mike Scott as president of Apple in early 1981, but by the end of 1982 he decided that he'd had enough of "curating high-maintenance egos" (Isaacson, 2011, p. 148). Steve Wozniak had taken extended leave from the company, and in any case would not have wanted the role. Steve Jobs understood that he was not the best person for the job, so he and Markkula started looking for someone outside the company who might be suitable. As a publicly quoted company, they needed someone who was known on Wall Street, and they also wanted someone who knew how to sell to 'ordinary' consumers. One person who fitted the bill was John Sculley, president of PepsiCo and responsible for the famous 'Pepsi Challenge' campaign. Jobs set about persuading Sculley to take the job with typical fervour. As negotiations came to a head, Sculley remembers:

> Steve had these deep, penetrating dark eyes, and he just stared right at me, probably 15 inches away, and said, "Do you want to sell sugared water for the rest of your life, or do you want to come with me and change the world?" It kind of knocked the wind out me, because no-one had ever said anything like that to me before. (Sculley, 2011)

Sculley took over as president of Apple in May 1983, some seven months before the Macintosh was launched. Markkula resumed his original role as chairman.

A month later, Wozniak also returned to Apple. He had left in February 1981 following a serious accident in the light aircraft that he had recently bought. The accident left him with temporary amnesia, but following his recovery a few months later, he decided to resume his education rather than return to Apple, and enrolled at UC Berkeley where he completed the degree course in electrical engineering and computer sciences that he had dropped out of some ten years earlier. He enrolled under the name 'Rocky Clark' (his dog was named Rocky Raccoon and he had recently married Candi Clark). During this period he also lost some $24 million financing the two US Festivals held in San Bernardino in September 1982 and May 1983, the first lasting three days and featuring The Ramones, The Kinks, Grateful Dead and Fleetwood Mac; the second running over four days with performances from The Clash, Van Halen, U2 and David Bowie (Wozniak & Smith, 2006).

Wozniak returned in a typically low-key fashion by simply walking into the building that housed the Apple II team and asking for a job. Sculley realised that Wozniak's return would boost the morale of the team, which had suffered in the face of the more glamorous Lisa and Macintosh projects, so he was welcomed back and put to work on the latest in the Apple II line, namely the Apple IIc (Linzmayer, 2004). Coincidentally, Microsoft co-founder Paul Allen was diagnosed with Hodgkin's Disease towards the end of 1982 and left the company in early 1983. His cancer was successfully treated but he did not return.

Initially the relationship between Sculley and Jobs was very good; however as the launch of the Macintosh approached, the tension between them grew. Sculley despaired of Jobs' impetuous and confrontational style; Jobs began to feel that Sculley was weak and easily manipulated. One early dispute was over the price of the Macintosh. Jobs wanted to keep it as low as possible and was pushing for $1,995. However Sculley insisted on adding a further $500 to cover the extravagant marketing push that Jobs had planned.

At launch, Apple was projecting worldwide sales of the Macintosh of 350,000 in its first year (Regis McKenna Public Relations, 1984), and indeed much of its design and manufacture was based on it being a mass-market, high-volume product. However the reality was that it was seriously hampered by its single floppy disk drive and just 128K of memory, which made it very slow and meant that users were continually swapping disks, while its price put it in competition with high-end PC compatibles. By the end of the year it was selling less than 10,000 units a month, but Jobs remained convinced that the future of the company lay with the Macintosh, and not with the Apple II. This was ignoring the fact that the Apple II was still selling like hot cakes: sales had reached one million in June 1983, and by October 1984 a further million had been sold. April 1984 had seen the arrival of the Apple IIc which, on launch day alone, took more orders than the Macintosh had managed in three months (Linzmayer, 2004).

The situation wasn't helped by the intense rivalry that Jobs encouraged between the two teams. Jobs had seized the Macintosh project from Raskin, a move that his team celebrated by hoisting a skull-and-crossbones outside the building in which they worked. Industry consultant Aaron Goldberg reports watching a high-spirited exchange in a local bar between the Macintosh team screaming, "We're the future!" and the Apple II team responding, "We're the money!" (Carlton, 1997, p. 13). Unfortunately, both were right.

By early 1985, Andy Hertzfeld and a number of other key members of the original Macintosh team had left, disillusioned with the way the company was going and confused by Jobs' increasingly erratic behaviour. Then, in February, Wozniak followed. As he later explained:

> At the time I was leaving, the people in the Apple II group were being treated as very unimportant by the rest of the company. This despite the fact that the Apple II was by far the largest-selling product in our company for ages, and would be for years to come. It had only recently been overtaken as number one in the world by the IBM PC, which had connections in the business world that we didn't have … If you worked in the Apple II division, you couldn't get the money you needed or the parts you needed in the same way you could if you worked in, say, the new Macintosh division. I thought that wasn't fair. (Wozniak & Smith, 2006, p. 265)

All of this was convincing Sculley that Jobs was not the right person to be running the Macintosh division. That April, the board of directors sided with Sculley, authorising him to remove Jobs from the Macintosh team. Jobs fought back, attempting to persuade them that it was Sculley that needed to go, but in a boardroom confrontation, it became obvious that he did not have their support. By the end of May, Jobs had been allocated the title of Chairman, but it was clearly a figurehead role with no real power. He formally resigned in September 1985 and over the following months sold all but one of his shares in Apple, making more than $100 million in the process. Together with a small number of colleagues who followed him from Apple, he founded a new company that he named NeXT with the intention of building high-end workstations for use in colleges and universities.

Talking about this period many years later, Sculley admitted:

> In hindsight, that was a terrible decision … I was focused on, 'How do we sell Apple II computers?' He was focused on, 'How do we change the world?' (Sculley, 2011)

In an interview with Robert X. Cringely, Jobs was less forgiving:

> What can I say? I hired the wrong guy … and he destroyed everything I spent ten years working for. (Jobs, 1996)

With Jobs gone, Sculley instigated a major reorganisation in an effort to dispel the poisonous atmosphere that had developed between the two teams. He put Jean-Louis Gassée, who had run Apple's highly successful French division, in charge of product development:

> I put Gassée in the job because I thought he was the closest thing to a Jobs ...
> Gassée became the cult leader. I was the outside spokesperson. (Carlton, 1997, p.
> 23)

Ironically, the Apple Macintosh was about to establish its niche in the market, and in the process, transform an industry. The first step was the Macintosh 512K, or 'Fat Mac', which was launched in September 1984 and overcame one of the Macintosh's main problems: its lack of memory. It was more expensive at $3,195, but it came with 512KB of RAM which allowed it to run more powerful programs. Then, in January 1985, while still leading the Macintosh team, Jobs announced Macintosh Office. Inspired by the networked systems that had been demonstrated at Xerox PARC, this was an ambitious plan to build something similar, and also to deliver a technology that would allow Macintoshes to work alongside IBM PCs in the business world.

There were four elements to Macintosh Office. First was the Macintosh 512K itself, which was reduced to $2,795. Then there was the AppleTalk Personal Network. All Macintoshes came equipped with the electronics necessary for a simple local area network, and for an extra $50 per Macintosh for the cable and the connectors, AppleTalk allowed you to connect up to 32 devices together. The plan was to introduce an expansion card that would allow IBM PCs and compatibles to be included, and there was also an intention that an AppleTalk network could be connected to the networking system that IBM was planning, although this had yet to materialise.

Another element was the 'file server'. This would be a machine designed solely to act as a file repository that anyone could access across the network. The concept eventually materialised two years later as AppleShare, a software package that could turn any Macintosh with sufficient memory into a file server. At the same time, Apple took the opportunity to launch the AppleTalk PC Card at $399 (Petrosky, 1987).

However the most important element was the Apple LaserWriter. This was a powerful laser printer that could print up to eight pages a minute at a resolution of 300 dots per inch. At $6,995 it was not cheap, but it could plug into an AppleTalk network, which meant it could be shared by more than 30 users. It also came with a Motorola 68000 processor connected to 512KB of ROM and 1.5MB of RAM, making it Apple's most powerful computer to date. As InfoWorld put it, "The

Apple LaserWriter might be described as a cross between a Macintosh computer and a Canon PC copier" (Bartimo & McCarthy, 1985).

The LaserWriter needed all this power so that it could run the interpreter for Adobe's page description language, PostScript. Named after the creek that ran behind their homes in Los Altos, Adobe had been founded in December 1982 by John Warnock and Charles Geschke, who had both worked at Xerox PARC in the late 1970s but, like so many of their colleagues, left because Xerox did not seem interested in turning their work into commercial products.

While at PARC, they had tackled the problems involved in generating and transferring the huge bitmaps that were necessary if you wanted to print a high-resolution image. At 300dpi, for example, approximately 1MB of memory is required to store a black-and-white image measuring 8.5 by 11 inches (standard American letter paper). The solution Warnock and Geschke came up with was to have the computer 'describe' the image it wanted to print in a language that they called Interpress. This description could then be sent across the network to the printer, which would turn it into the dots necessary to render the image on the paper. For example, the computer might say, "draw me a solid black rectangle with one corner at coordinates 100, 100 and another at 200, 350." The text of such an instruction could be expressed in less than a hundred bytes, but once rendered could end up as a bitmap requiring several hundred kilobytes to store.

One of the advantages of using a page description language such as Interpress is that far less data needs to travel between the computer and the printer. The other is that the instructions are independent of the resolution of the target device. A low-resolution printer might render a delicate design quite crudely, while a high-resolution laser printer would be able to render it in much more detail from the same data. The rendering itself is done by what amounts to a computer within the printer which is running a program that converts the instructions into a bitmap at the appropriate resolution.

At Adobe, Warnock and Geschke refined the language they had invented into PostScript, and also met up with Steve Jobs:

> We met Steve Jobs about three months after we started Adobe. He called us and said: "I hear you guys are doing great things – can we meet?" He came over to our tiny office in Mountain View and saw the early stages of PostScript. He got the concept immediately and we started about five months of negotiations over our first contract. Apple invested $2.5 million into Adobe and gave us an advance on royalties. This allowed us to help Apple build the first LaserWriter.

> Without Steve's vision and incredible willingness to take risk, Adobe would not
> be what it is today. (Narayen, 2011)

One of the features of PostScript was its ability to reproduce text in a wide variety of typefaces. This was achieved through small 'font' files that defined the shape of each character within a typeface, allowing you to write instructions like "Print the following text in 24pt Helvetica" and have the LaserWriter translate it onto the page in beautifully flowing script. The LaserWriter came with definition files for the Times, Helvetica and Courier typefaces built in, and others could be loaded into a Macintosh and then copied across the network into the LaserWriter's memory. Later that year, Apple announced that it had licensed a further 35 typefaces from Mergenthaler and from ITC (International Typeface Corporation), including Palatino, Optima, ITC Souvenir and ITC Zapf Dingbats (which instead of letters and numerals, defined a range of symbols).

While at college, Jobs had attended a class on calligraphy which had left a lasting impression. He later suggested, "If I had never dropped in on that single course in college, the Mac would have never had multiple typefaces or proportionally spaced fonts." He continued, "And since Windows just copied the Mac, it's likely that no personal computer would have them" (Isaacson, 2011, p. 41). An unlikely state of affairs, but it demonstrated how bitter he felt.

As part of the Macintosh Office announcement, Jobs took the opportunity to introduce a number of supporting products that had been developed by other companies. The most important was the PageMaker application from Aldus Corporation. Its founder Paul Brainerd understood publishing, having worked on the *Star Tribune* and then for Atex Publishing Systems which had developed a publishing system for minicomputers that was widely used in the newspaper industry. Brainerd was based at Atex's research and manufacturing centre in Redmond, just outside Seattle, where he rose to vice-president, only to leave in 1984 when Kodak bought the company and closed the centre down.

Shortly after leaving Atex, Brainerd was introduced to Apple and the work they were doing on the LaserWriter and Macintosh Office. Immediately seeing its potential, he employed a number of his colleagues from Atex and set about developing software that would turn Macintosh Office into what he called a "desktop publishing" system (Mace, 1985). He named his company in honour of Aldus Manutius the Elder, a 15th century printer and publisher from Venice who is credited with inventing not only italic text but also the semicolon. Launched in July

1985, Aldus Pagemaker displayed a facsimile of a printed page on the Macintosh screen, accurately reproducing multi-column type in the supported typefaces, together with images and other graphics that could be positioned and edited using the Macintosh's intuitive point-and-click interface. The page could then be printed on a LaserWriter at a resolution of 300dpi.

At the time, most professional journalists and writers were still using typewriters. Their finished copy, together with any accompanying photographs or illustrations, would be sent to a typesetter, a specialist company with the equipment and skill to turn them into a form that a printing press could handle. A couple of days later, the typesetter would return copies of the text set in the required font and width (known as 'galleys') which a layout artist could cut and paste into a representation of the desired page layout using a ruler, pencil and knife. This 'paste-up' would then be returned to the typesetter, where it was used as a template for creating the final plate ready for printing.

Most typesetters used a photographic process although some of the more traditional newspapers were still using a 'hot metal' process that involved casting text into lead-based alloy 'slugs' ready for inking and printing. Some typesetters were using computer systems, made by companies such as Atex or Compugraphic, but the equipment was highly specialist and extremely costly. Some could even accept copy sent in on a floppy disk, allowing writers to use word processing software running on perhaps a PC compatible or an Apple II. However the process remained involved and highly partitioned by the respective skills required.

In the UK particularly, the situation was further complicated by unionisation. Most journalists belonged to the National Union of Journalists (NUJ) while their fellow layout artists, and those working in the typesetting companies, belonged to the National Graphical Association (NGA). Most of the less skilled print workers were members of the Society of Graphic and Allied Trades (SOGAT). The trade union movement in Britain had become highly politicised following the 1978 'Winter of Discontent', and restrictive practices were common. Publishers were becoming frustrated at the expense such practices caused, and what they perceived as a reticence to accept new technology. With the election of Margaret Thatcher, and the legislation which her government introduced to curb the power of the trade unions, change was inevitable.

Eddie Shah, owner of the Warrington-based Messenger Group, which published a number of local newspapers, was the first to use the new legislation in

1983, forcing the NGA to back down and allowing him to employ non-union members. He was followed by Rupert Murdoch whose company News International owned *The Sun* and *News of the World*, and in 1981 had acquired *The Times* and *The Sunday Times*. Murdoch had spent a huge amount of money buying up the American television stations that would form the basis of his Fox Broadcasting Company, and needed to increase the revenue he was getting from his British newspapers, which were all based in Fleet Street. That meant cutting jobs and introducing more efficient technology. He therefore embarked on a strategy that many claim was a deliberate attempt to provoke the unions into a confrontation that could only lead to their demise.

His first step was to build a brand new high-tech plant in Wapping, based around Atex equipment. Although ostensibly to produce a new evening newspaper called *The London Post*, the plant incorporated many high security features and quickly became known as 'Fortress Wapping'. In January 1986, following protracted union negotiations, the new plant was used to print *The Sunday Times* supplement, which resulted in strike action from some 6,000 union members. Murdoch's reaction was to fire the striking workers and take legal action against the SOGAT and NGA unions. He then transferred production of all four of his Fleet Street newspapers to the new plant. As the dispute escalated there were repeated clashes between the police and union pickets, often violent, until the unions finally abandoned the fight more than a year later.

It was against this background that Apple's desktop publishing system was launched, at a stroke slashing the cost of the equipment needed to create a professional publication from several hundred thousand dollars to perhaps ten or twelve thousand (around $23,000 in today's prices). Furthermore, it redefined the skills required to bring a publication to print. Graphic artists risked losing their livelihood unless they could master a desktop publishing application such as PageMaker, while much of the work traditionally carried out by typesetters was no longer necessary, as it was carried out automatically by the software.

In an effort to establish PostScript as a standard, Adobe was quick to license it to other printer manufacturers, with the result that it soon became possible to connect a Macintosh system directly to an imagesetter such as the Linotronic 100 or 300, capable of printing at resolutions comparable to the equipment used by the traditional typesetter, but at a fraction of the cost. Service bureaux appeared that could output your designs from a floppy disk to a LaserWriter for a few pence a page, or to high resolution film ready for printing for a couple of pounds. Many of these were typesetting companies which had realised that survival meant embracing the new technology. In the words of *InfoWorld* reporter John Gantz:

> Electronic in-house publishing is very exciting. It could offer tremendous cost savings, new methods of database delivery, and wrenching career decisions for graphic artists.

However he warned:

> But, as personal computers bring layout and design capability to the masses, excitement will be accompanied by an explosion of god-awful art. Do you think that people can't spell? Wait until you see their text/graphics page layouts, when they have a million fonts and point sizes at their command. As Times Square is to neon lights, 'personal' electronic publishing will be to page design. (Gantz, 1985)

His fears were not unfounded, but the Macintosh quickly proved that it could also produce results every bit as professional as traditional technology. In 1986, working with author MacKinnon Simpson, Robert Goodman published *WhaleSong*, a full-colour coffee-table book chronicling the history of whaling in Hawaii. Simpson designed the book using Aldus PageMaker running on the Macintosh Plus, which came with 1MB of memory and the connections necessary to plug in to a 20MB hard disk drive. This was still a monochrome system, so Goodman used a local company to handle the colour production, but as he said, "I probably saved between $40,000 and $60,000 in production costs. Without the computer, there was no way I could have done this book." (Banks, 1987). For the second edition, published three years later, they used a Macintosh II with a colour display, an early colour scanner from Sharp and a pre-production version of Adobe Photoshop to accomplish all the colour work themselves, on their own desktops.

For graphic designers and those working in the art departments of publishing or marketing companies, the Macintosh had become indispensable, and it was also beginning to find its place within the music and film industry. However these were niche markets. The vast majority, while they liked the idea, could not justify the cost. The PC compatible with its word processors, spreadsheets and accounting packages could do what was needed, albeit with rather less style.

11.   The battle of the WIMPS

Although announced in November 1983, it was to be another two years before Microsoft Windows actually became available. In the interim, several other 'windowing' systems appeared. VisiCorp released Visi On, IBM's TopView came out in August 1984 and was closely followed by Tandy's DeskMate which ran under MS-DOS on the PC compatible Tandy 1000. However by far the most well-known was Digital Research's Graphical Environment Manager, or GEM for short.

Having been trumped by Microsoft on the IBM PC deal, Digital Research had returned to earlier work that Tom Rolander had done with MP/M (Multi-Programming Monitor for Microcomputers) in an effort to produce a multi-user version of the CP/M-86 operating system. Rolander had in fact demonstrated MP/M to IBM's representatives at the fateful 1980 meeting, as he felt that it was better suited to the IBM PC than CP/M, but they had shown little interest, because they were looking for something less ambitious (Rolander, 2007).

The result was Concurrent DOS 3.1 which Digital Research offered to hardware manufacturers in March 1984 as an alternative to MS-DOS. In addition to programs written for CP/M-86, Concurrent DOS 3.1 supported a 'PC-Mode' with the claim that, "Most of the popular software developed for PC-DOS runs under the new release of Concurrent DOS" (Digital Research, 1984). Furthermore, users could run up to four applications at the same time, switching between them from the keyboard and even displaying them simultaneously within windows that could be moved around the screen.

Although impressive, Concurrent DOS proved to be too much, too soon. As one commentator put it, "It's like turning a passenger car into a dumptruck" (Cook, 1984). At the time, most desktop computers came with just 64 or 128KB of memory which was barely enough to run one program, let alone four, and adding more was a costly exercise. Furthermore, PC-Mode proved incompatible with many important DOS applications.

However Concurrent DOS did come with GSX (Graphics System eXtension), a library of graphics routines written by a team led by Lee Jay Lorenzen, another renegade from Xerox PARC who had worked on the user interface for the Star.

Building on these routines, GEM (Graphical Environment Manager) was designed to run on top of an operating system, which could be MS-DOS or Digital Research's own Concurrent DOS, and present a graphical user interface similar to that of the Apple Macintosh or the Xerox Star, complete with overlapping windows, icons and drop-down menus, and controlled by a mouse. Instead of typing arcane commands, the user could open files and run programs by double-clicking the appropriate icon. If the program was a DOS application such as WordStar or Lotus 1-2-3, it would take over the screen, returning you to GEM Desktop when you exited (although some would dump you back at the DOS prompt, from where you would have to reload GEM manually). Programs that had been written specifically for GEM could be run within a window and controlled by the mouse. These included GEM Write, GEM Paint and GEM Draw.

Digital Research demonstrated GEM at the Comdex computer exhibition towards the end of 1984 and released a version for PC compatibles and other non-compatible machines based on Intel's 8086 processor, such as the Apricot F1, in February 1985. At around the same time, GEM was revealed as the user interface for a brand new machine from Atari, which was now owned by Jack Tramiel.

In January 1984 Tramiel left Commodore, the company he had founded 26 years earlier, following a falling out with its principal investor, Irving Gould, over the way in which the company should be run. In the meantime Atari had fallen on hard times, partly as a result of competition from Commodore, so Tramiel entered into negotiations with Atari's owner, Warner Communications, and secured its purchase for a knock-down price that July. Once in charge, he concentrated resources on the new machine. The Atari 520ST was announced in January 1985 at the Winter Consumer Electronics Show (CES) in Las Vegas and became generally available in June.

Like the Apple Macintosh, the Atari 520ST used Motorola's 68000 processor but came with a more than adequate 512K of RAM for just $800 (£749 in the UK) including a 3.5-inch floppy disk drive and monochrome monitor. It was even available with a colour display for only $200 more – something that wasn't to appear for the Macintosh for another two years. Sitting alongside the RAM was a further 192K of ROM which stored both the operating system and GEM itself, which meant that they were loaded automatically as soon as you switched the machine on, leaving the full 512K available for your applications. The operating system was based on GEMDOS, a project that Digital Research had been working on to build something specifically designed to support GEM. The combination of GEM and GEMDOS was dubbed TOS, which originally stood for The Operating System but soon became Tramiel Operating System in honour of the company's new owner. Likewise, the Atari ST quickly became known as the 'Jackintosh'.

Then, a month later, Commodore announced something even more powerful. The Amiga had first seen the light of day a year earlier when the fledgling company Amiga Corporation demonstrated a prototype at the Summer CES in Chicago. The prototype (codenamed 'Lorraine') used the same Motorola 68000 processor as the Atari ST and the Apple Macintosh, but was assisted by three very advanced special-purpose chips. Dubbed Agnus, Denise and Paula, these had been developed by a team headed by Jay Miner, who had previously worked at Atari where he had been responsible for the graphics processors used by Atari's earlier games consoles and home computers. The result was a machine capable of unprecedented graphic animation, able to display up to 4,096 colours on screen at once, and of synthesising up to four simultaneous sounds in stereo (Mace, 1984).

However Amiga was running short of funds, and in November 1983 had entered into an agreement with Atari that gave Miner's former employee access to Amiga's technology. As Miner described it:

> Atari gave us $500,000 with the stipulation that we had one month to come to a deal with them about the future of the Amiga chipset or pay them back, or they got the rights. This was a dumb thing to agree to but there was no choice. (Miner, 1992)

As it became apparent that Atari was itself in trouble, Amiga started looking for a way out and entered into negotiations with Commodore which resulted in Commodore purchasing the company outright in August 1984 and repaying the money owed to its arch-rival Atari. This was just one month after Tramiel had taken charge of Atari, and by the following week, Commodore was facing a lawsuit claiming breach of contract and seeking damages of $100 million (The Modesto Bee, 1984).

The suit was eventually settled out of court some two years later. Meanwhile, in July 1985 – six months after the announcement of the Atari ST – Commodore went ahead and launched the Amiga at the CES in Chicago with the help of the artist Andy Warhol and singer Debbie Harry. Warhol used a digital camera to take a snapshot of the singer's face which he then manipulated in real time on the computer screen. Asked, "What other computers have you worked on before?" Warhol replied, "Oh, I haven't worked on anything. I've been waiting for this one." (Branwyn, 2010).

The Amiga was more expensive than the ST, costing $1,295 for a version with 256K of memory and a floppy disk drive that could be connected to a colour television, or $1,790 complete with colour monitor. However this was an extraordinary machine. Not only did it offer unprecedented graphics and sound; it also came with AmigaOS, a multi-tasking operating system that presented a mouse driven graphic user interface with windows, icons and drop-down menus, similar to GEM or the Macintosh but in full colour. An early demonstration had a document being edited in a foreground window while a multi-coloured ball bounced smoothly up and down in the background, complete with sound effects as it collided with the edges of the screen. Such a performance would not be possible from a PC compatible for nearly a decade, and yet the machine cost little over half the price of a Macintosh.

Despite their technical prowess, the Amiga and the ST were to have problems finding their place in the market. Both Commodore and Atari were well known to gamers and hobbyists, but this made it difficult for them to enter the business world, as corporate buyers were reluctant to buy into something that carried the same badge as the computer their kids played with at home. On the other hand, they were too expensive to appeal to most home users. There was also the problem of support: the Apple Macintosh succeeded as a desktop publishing system only in combination with Aldus PageMaker and the LaserWriter: the Amiga might be able to produce wonderful 3D colour charts and graphs, but these weren't much use in the boardroom without high-quality colour printers and presentation software that could work with their existing systems.

The Atari ST did find a niche in the music industry, largely thanks to its built in MIDI connector which could be used to control professional music synthesisers. The composition applications Cubase and Logic Pro, both still widely used, were originally written for the ST before they became available for the PC and the Macintosh. The Amiga was used in the early 1990s to create special effects in TV and film, most notably for the TV series *SeaQuest DSV* and *Babylon 5*. Both machines also succeeded in the games market that they knew so well – particularly once Commodore released the cheaper Amiga 500 in January 1987 at $699 without a monitor (£499 in the UK). All in all, it is estimated that some four million Amiga computers had been sold worldwide by the time the line was discontinued in 1993 (Knight, 2006). The Atari ST was not as popular, but did sell well in Europe. That said, neither machine accounted for more than three or perhaps four per cent of the total number of personal computers sold.

For Apple, 1985 was proving to be a difficult year. Against a background of increasing conflict in the boardroom, which would result in Steve Jobs leaving the company, Apple was facing possible competition from Microsoft with Windows, and in the meantime Digital Research had launched GEM which turned a PC into something that looked dangerously like to a Macintosh. Apple was not particularly concerned about the use of GEM on the Atari ST, but it was worried about the possibility of GEM being bundled with the IBM PC or the PCjr. Shortly after GEM was launched, Digital Research's marketing director Thomas Byers was quoted as saying:

> Essentially what we've done is say that we will provide the GEM engine to the OEMs [Original Equipment Manufacturers] across the world – Atari, Commodore, even Apple if they want it. It will be on their machines when you buy the computer, and your GEM application will run. The software developer who is concerned about the IBM channel will be able to include a disk that has GEM on it which will allow their application to run on any of the IBM PC family, including the Junior all the way up to the AT.

Asked whether IBM would be licensing GEM, Byers continued:

> "IBM is a large OEM customer of Digital Research. We did more business with IBM last year than we did with any other hardware manufacturer, believe it or not. That is not to say it was GEM or it wasn't GEM; I couldn't say that even if I wanted to. They do their own announcements. (Bateman, 1985)

Apple's response was to threaten Digital Research with legal action, claiming that GEM breached Apple copyrights – a claim that could scupper any deal that IBM might have been contemplating. Digital Research backed down, agreeing to make changes to the GEM user interface, and to pay Apple an undisclosed amount (New York Times, 1985). When GEM 2.0 was released the following March, the PC version came without the trash-can icon and instead of multiple overlapping windows it could only display two fixed windows, one above the other. GEM on the Atari ST continued unchanged.

The situation with Microsoft was more complicated. Bill Gates was very keen on the Macintosh and Apple was glad of his support. Gates had participated in the Macintosh launch with the statement, "In the software business volume is everything: you want to be able to sell onto a large set of machines. Microsoft is choosing this Apple Macintosh environment because over time the other environments won't be interesting." In October 1983, during a rather zany presentation hosted by Jobs under the title Macintosh Software Dating Game, Gates stated to considerable applause that "during 1984, Microsoft expects to get half of its revenues from Macintosh software" (Gates B. , 1983). A bit optimistic but in January 1985, Microsoft released Microsoft Word for Macintosh. Microsoft Excel for Macintosh was announced in May 1984 but was not released until September of the following year.

For Microsoft, the Macintosh represented an opportunity to hedge the very considerable commitment it had made to IBM, but only if it sold in quantity. By mid-1985 it was apparent that the Macintosh was not selling as well as was hoped and it occurred to Jeff Raikes, who had worked at Apple and was now on Microsoft's Macintosh team, that Apple could benefit by licensing Macintosh technology to other manufacturers, so establishing it as a standard and putting it on a similar footing to the PC compatible. Raikes put the idea to Gates and in June 1985, already aware that Jobs was no longer in charge, Gates sent a confidential letter outlining the strategy to John Sculley and Jean-Louis Gassée. The letter stated:

> As the independent investment in a 'standard' architecture grows, so does the momentum for that architecture. The industry has reached the point where it is now impossible for Apple to create a standard out of their innovative technology without support from, and the resulting credibility of other personal computer manufacturers. Thus, Apple must open the Macintosh architecture to have the independent support required to gain momentum and establish a standard ...
>
> Apple should license Macintosh technology to 3-5 significant manufacturers for the development of 'Mac Compatibles' ... Microsoft is very willing to help Apple implement this strategy. We are familiar with the key manufacturers, their strategies and strengths. We also have a great deal of experience in OEMing system software... (Carlton, 1997, p. 40)

The manufacturers that Gates suggested included AT&T, DEC, Hewlett-Packard, Olivetti, Siemens, Texas Instruments and even Xerox.

In hindsight, such a strategy would seem detrimental to Microsoft's interests, giving hardware manufacturers the choice between building a PC compatible and licensing Windows, or building a Macintosh compatible and licensing Mac OS from Apple, which would put the two companies in direct competition. However, at the time the letter was written, Windows had yet to appear and its success was far from certain. Gates saw the strategy as something that could increase the market for his new Macintosh applications and reduce his dependency on IBM.

If Apple had taken this route, the history of the personal computer would have been rather different. However Gassée in particular was having none of it. All he could see was falling revenues if Apple opened itself to competition from clone manufacturers, leaving the company with less to invest in the technologies that it needed to keep ahead. As he later stated:

> You can't be in both the hardware and the licensing businesses at the same time ... IBM licensed key parts of the original PC design and, for its reward, lost the PC market in spite of its effort to regain control with a new bus architecture ... and a new software platform ... (Gassée, 2010)

Meanwhile, Microsoft Windows was slowly coming to fruition. Gates had been quite open with Jobs and Sculley about his plans for Windows, which was making Sculley increasingly nervous. As with Digital Research earlier, Sculley's response was to threaten Gates with legal action for copyright infringement. However Gates was not impressed – as far as he was concerned, both Windows and the Macintosh were derived from work originally carried out by Xerox and, furthermore, he was already making a considerable investment in the Macintosh through Word and Excel. Aware that Apple needed these applications if the Macintosh was to make an impression with business buyers, Sculley agreed to a licensing agreement with Microsoft that was eventually signed just a few days after Windows was launched:

> The parties have a long history of cooperation and trust and wish to maintain that mutual beneficial relationship. However, a dispute has arisen concerning the ownership of and possible copyright infringement as to certain visual displays generated by several Microsoft software products ...
>
> For purposes of resolving this dispute and in consideration of the license grant from Apple ... Microsoft acknowledges that the visual displays in [Excel, Windows, Word and Multiplan] are derivative works of the visual displays generated by Apple's Lisa and Macintosh graphic user interface programs ...
>
> Apple hereby grants to Microsoft a non-exclusive, worldwide, royalty-free, perpetual, non-transferable license to use these derivative works in present and future software programs and to license them to and through third parties for use in their software programs. (Apple v. Microsoft, 1994)

In return, Microsoft granted a similar licence for Apple to use "any new visual displays created by Microsoft" for Windows over the following five years, and agreed to delay the release of its Excel spreadsheet application for computers other than the Macintosh.

Microsoft Windows 1.0 was finally released in November 1985. Essentially this was a colour version of GEM with pop-up dialogue panels and drop-down menus that could all be controlled with a mouse. Disk drives were displayed as icons but individual files were listed by name, although they could still be opened by pointing and clicking. As in the original demonstration two years earlier, the windows themselves could be tiled but not overlapped. Also provided were a number of applications and utilities that had been specially written for Windows and opened into windows that had their own specific drop-down menus. These included Windows Write, Windows Paint, Calendar, Cardfile, Notepad, Clock, Calculator, Terminal and the game Reversi.

Despite costing just $99 (£86 in the UK), Windows 1.0 had little impact. Few PC users had machines powerful enough to take full advantage of what it had to offer, so most carried on running their applications directly from MS-DOS. Nevertheless, Windows 1.0 did have a number of features that would come into their own as the years progressed. For a start, it allowed multiple programs, including conventional DOS applications, to be loaded at the same time. Only one program could be active at a time, but users could switch between Lotus 1-2-3, Microsoft Word and Windows Paint, for example, without having to exit one before loading up the next. It also came with a utility called Clipboard that allowed you to select data from one application and paste it into another, so you could copy an image that you had created in Windows Paint and paste it into a document that you were preparing in Microsoft Word.

Windows 1.0 was itself an MS-DOS program, which meant that active programs had to fit alongside its own program code within the 640K of memory that the operating system allocated to applications. However Windows could 'swap out' inactive programs to a hard disk. Furthermore, it came with a utility called RAMDrive which could fool MS-DOS into seeing any additional memory, above and beyond 640K, as though it was a very fast disk drive. Memory expansion cards such as the Intel Above Board were very expensive but could add an extra 2MB to your system. The technology was flaky in this first version, but when it worked you could switch between multiple applications in a reasonably smooth fashion. It all added up to little more than the Macintosh had been capable of doing since its launch nearly two years ago, but it wasn't the Macintosh that was sitting on most people's desks.

By the middle of 1986 the trade magazine *InfoWorld* was reporting that "An increasing number of software developers ... are leaning to Microsoft's Windows over rival operating environments ..." and quoting Paul Brainerd of Aldus, which had tested prototypes of Aldus PageMaker running under both Windows and GEM: "Our conclusion was that Windows was the clear winner in that context from both a technical and a marketing point of view." That said, the ordinary user could find little reason to get excited. Ray Potts, in charge of personal computer purchasing at Wells Fargo Bank of San Francisco, was quoted in the article as stating: "Way less than one per cent of our micro users have so far requested these types of products."

That August, Microsoft released Windows 1.03, a minor update that added a PostScript printer driver so that Windows applications could print to printers such as the Apple LaserWriter that supported PostScript. In a vote of confidence, Aldus released PageMaker for the PC 1.0, a version of the popular desktop publishing application that could run on PC compatibles, in March 1987. For users who had not already installed Windows 1.03 itself, the program came with a Windows 1.03 'runtime': a subset that provided just those portions of Windows code that PageMaker needed to run. Microsoft Excel 2.05 for Windows followed in October. Such programs ran painfully slowly, even on a top-end machine, but reasons to buy Windows were increasing.

Meanwhile Microsoft was looking beyond the limitations of the 8086 processor towards a future that lay with the new modes of operation made possible by the 80286, which drove IBM's PC AT and a growing number of PC compatibles, and the 80386 which Intel had just launched. Bill Gates was not convinced by IBM's strategy, but IBM's still powerful position in the market, and the substantial revenue stream that Microsoft derived from PC-DOS, argued against jeopardising the relationship – at least for the time being. This was one of the reasons why he had signed a Joint Development Agreement with IBM in August 1985, "in order to establish a working relationship between IBM and MS for evaluating the feasibility of and/or developing system software products based upon IBM PC DOS and/or

MS DOS ..." which would take advantage of these more advanced processors.

The fruit of this agreement was made public in April 1987 through a number of announcements from both companies. On IBM's side there were the PS/2 machines with which it hoped to establish a new industry standard and wrest the initiative from the 'Gang of Nine' (see page 98). Then there was OS/2, the new operating system that had been jointly developed under the agreement. Finally there was Microsoft Windows 2.0, which introduced a number of improvements to the way in which Windows managed memory and a number of enhancements to the graphic user interface – including the ability to overlap windows in a fashion more reminiscent of the Apple Macintosh. It was also made clear that Windows 2.0 was to be seen as a precursor of Presentation Manager, the graphic user interface for OS/2 which would be introduced at a later date.

The first version of the new operating system was OS/2 Standard Edition 1.0 which was released that November at a price of $325. This was a text-based system that looked much like MS-DOS, and indeed shared many of the same commands. However under the hood it ran in the protected mode supported by the 80286 and 80386, which meant that it could directly access 16MB of memory (as opposed to the 1MB available to MS-DOS) and could support more than one application running at the same time in a far more stable fashion than was possible on an 8086-based machine. The catch was that it could not do this with conventional DOS applications – this feature only worked with applications that had been specifically written for OS/2.

Despite the promise of an OS/2 version of Microsoft Excel and an announcement from Lotus that it would produce OS/2 versions of a number of its programs, including the popular 1-2-3 spreadsheet application, OS/2 did not garner much support. This first version could run on machines that used the 80386, but it was designed for the 80286, which severely hindered its ability to run the DOS-based programs that most people used. Without the Virtual 8086 mode offered by the 80386 processor, OS/2 1.0 could only run one DOS application at a time, by switching the 80286 back and forth between real mode and protected mode, which the processor was not designed to do. The result was cumbersome, but IBM was not unduly concerned. To quote the launch press release:

> It is anticipated that usage of the DOS environment will diminish as many applications are converted to IBM Operating System/2 applications to obtain the advantages of larger memory and multi-programming. The IBM Operating

> System/2 DOS environment preserves the end user's existing software investment
> during the migration to running only IBM Operating System/2 applications.
> (IBM, 1987)

IBM's William Lowe was quoted as saying that he expected it would take three years at the most for OS/2 to become the dominant operating system for the PC (O'Reilly, 1987).

IBM and Microsoft continued to present a united front, attempting to position Windows as a stop-gap best suited to smaller businesses, and OS/2 as a high-end system that would eventually replace it. However, behind the scenes, the two companies were already moving in different directions. In May 1988, Microsoft released Windows 2.10 in two versions, one optimised for the 80286 processor and the other for the 80386. Both were designed not only to run programs specifically written for Windows, but also to provide a useful environment for running DOS applications. Both cost considerably less than OS/2, at just $99 each.

Despite its name, Windows/286 operated in real mode and could run on a conventional PC compatible with an 8086 processor. However, when run on an 80286 machine, it made use of a quirk in the way in which the processor worked which effectively added an extra 64KB to the 1MB accessible to the 8086. Windows/286 could store part of itself in this High Memory Area (HMA), so giving applications a little more memory in which to run.

Windows/386 was rather more revolutionary, in that it was the first version of Windows to make use of the Virtual 86 mode of the 80386 processor. As a result it could multi-task both DOS and Windows applications in a far more stable and manageable fashion. DOS applications could be displayed in overlapping windows, and if multiple applications wanted to access a printer or communications port at the same time, Windows/386 would pop up a dialogue panel so that you could choose between them.

A few months later, the companies jointly released OS/2 Extended Edition 1.0 at a retail price of $795. Like the Standard Edition, this was designed for the 80286 processor and fell far short of Windows/386 in its support for DOS applications. However it did come with a powerful database manager and facilities for communicating with IBM mainframe and minicomputers that were in keeping with IBM's plans to integrate the PS/2 range into its overall corporate strategy. This was followed in November by OS/2 Standard Edition 1.1, which marked the appearance of the Presentation Manager graphical user interface. Extended Edition

1.1 arrived a few months later and OS/2 1.2, which introduced a much improved system for managing disk storage, was announced in May 1989.

By now it was becoming increasingly obvious that Microsoft and IBM were parting company. Nevertheless the two companies maintained the pretence of a united front, announcing in November 1989 that their proposed "platform for the '90s" would be based solely on OS/2. The joint press release explicitly stated:

> OS/2 1.2 is recommended for systems with at least 3MB of memory ... Both companies are making a concerted effort to enable OS/2 for 2MB entry systems. Customers should [only] plan to use Microsoft Windows to implement graphic applications on platforms with less than 2MB of memory. (Microsoft Limited, 1989)

The dichotomy was particularly unnerving for software companies. Anyone developing an application for this brave new world had to choose whether to target Windows, OS/2 or Presentation Manager, and whichever they chose involved a steep learning curve and a big risk. On the one hand, they were being encouraged to focus on OS/2 and Presentation Manager, but on the other, an increasing number of their customers were using Windows, primarily because Windows applications were so much easier to use than the DOS applications they were used to. This made Windows particularly attractive to corporate IT managers, who might be responsible for training and supporting hundreds of users at a time.

Then, in May 1990, Microsoft launched Windows 3.0 through a series of coordinated high-profile events across the world, and the game was up. Sales of Windows had been steadily increasing, to the extent that by the end of 1988, Microsoft was shipping around 50,000 copies a month (LaPlante, 1990). However more than 100,000 copies of Windows 3.0 were sold within a fortnight of its launch (The New York Times, 1990), and nearly three million by the end of its first year. By contrast, total sales of OS/2 since its launch more than three years previously amounted to some 300,000 (Parker, 1991).

Windows 3.0 was the right product at the right time. For a start, it was designed to make much better use of Intel's 80386 processor at a time when machines based on the 80386 were becoming affordable. You could buy an Amstrad PC2386, for example, for less than £2,000 complete with 4MB of memory, a 65MB hard disk and a colour display. Processor power had steadily increased with each passing year: the 20MHz clock driving the PC2386's processor made it capable of executing more than six million instructions per second; six times as many as the less efficient

6MHz 80286 used by the PC AT and 18 times as many as the 8086 inside the original IBM PC. Windows 3.0 made much better use of large amounts of memory too, which meant that it felt more responsive.

Although it really came into its own on an 80386-based machine, Windows 3.0 was specifically designed to work on 80286 and even 8086-based PCs as well, automatically checking your hardware as it was installed and configuring itself to make the best use of the processor and memory available. It was also much better at running ordinary DOS applications which, together with its cost of $149 (or just $50 to upgrade from an earlier version), was a sharp contrast with OS/2 and Presentation Manager. With Windows 3.0, Microsoft was catering for the needs of the ordinary PC user, while OS/2 seemed more concerned with a corporate strategy of IBM that was becoming increasingly irrelevant. As one contemporary commentator put it:

> Windows 3.0 is more than an update; in many respects, it's a whole new environment that will provide users with many of the features they've wanted for years but thought they'd have to switch to OS/2 or Unix to get ... As a result, many users will now be able to use Windows as their primary operating environment, launching and running multiple Windows-based and traditional DOS-based applications. (Miller, 1990)

The other big factor that boosted its adoption was the number of Windows applications that were actually available – applications that displayed the Windows user interface and took advantage of the additional memory available when operating in protected mode. In addition to Microsoft's own Word for Windows and Excel, there were Micrografx Designer, Corel Draw and Adobe Illustrator which, like Aldus PageMaker, was already popular on the Apple Macintosh. Adobe released a version of Adobe Type Manager, which gave Windows 3.0 users access to PostScript fonts, while both Lotus and WordPerfect announced that they were developing Windows 3.0 versions of their popular spreadsheet and word processing applications. Then, in November, Microsoft consolidated its position with the launch of Office 1.0 for Windows which combined Word, Excel and its office presentation application PowerPoint into a single package. Anyone could now store data and generate graphs that could be inserted into professional-looking documents and presentations without leaving an environment owned entirely by Microsoft.

148

Microsoft had another card up its sleeve, too. Back in 1988, a leading software engineer named David Cutler decided to leave Digital Equipment Corporation (DEC), where he had been responsible for the design of a number of operating systems for minicomputers such as the PDP-11 and its successor the VAX-11. Gates learned of Cutler's intentions and approached him to see if he would be interested in joining Microsoft. Cutler agreed on condition that he could bring a number of colleagues, and so that November a small but highly talented group of programmers was brought together to start work on a future 32-bit version of OS/2 that would be more 'portable', in that it wouldn't be tightly matched to one particular platform but instead could be easily tailored to work on a variety of microprocessors from different manufacturers. The initial system was designed around a new 32-bit processor that Intel was developing under the codename 'N-Ten', and so the project became known as 'NT'. The processor didn't work out, but by early 1990 the team had it running on a different 32-bit processor from MIPS Technologies, and started work on a version for the much more popular Intel 80386.

An important part of any operating system is its Application Programming Interface (API), a set of instructions that an application can use when it needs to interact with the operating system, perhaps to ask it to save something to disk or display something on screen. With the launch of Windows 3.0, and the acclaim that followed, it occurred to the NT team that they could just as easily create an API for their new operating system that was based on that of Windows 3.0, rather than OS/2. Such an API wouldn't be able to run Windows applications directly, but it would make it much easier for software companies to create versions of their Windows applications that would run on NT. Yes, they were supposed to be creating the next version of OS/2, but in the meantime they set about developing a 32-bit version of the 16-bit Windows 3.0 API, which they called Win32.

Microsoft loved the idea, but IBM weren't so keen. Mark Lucovsky, a member of the NT team, remembers IBM's reaction when first presented with the idea:

> At first, they thought Win32 was a fancy name for OS/2. Then you could just see it on their faces: "Wait a second, this isn't OS/2." (Thurrott, 2003)

The balance of power had changed. That September, the two companies announced a 'realignment' of their relationship such that IBM would take over responsibility for the development of OS/2 2.0 and Presentation Manager, leaving Microsoft to concentrate on Windows and what was still being called OS/2 3.0.

Asked to describe the new relationship between the two companies, one Microsoft executive pleaded with *InfoWorld*, "Say anything, just don't call it a divorce" (Parker, 1990).

But that's exactly what it was. Microsoft had been describing NT as a core component of OS/2 3.0, but in July 1991, Microsoft's Steve Ballmer made it clear that the new operating system would be called Windows NT (which now stood for New Technology), would present a user interface similar to the forthcoming Windows 3.1, and would run 32-bit Windows applications. He added, "The OS/2 API is something we could ship as a separate module to help OS/2 customers come back to Windows NT" (Johnston, 1991). Needless to say, no one expected that to actually happen.

Windows 3.0 was followed in 1992 by Windows 3.1, which incorporated improvements based on the feedback of thousands of both domestic and corporate users. As Microsoft put it, "User suggestions were the driving force behind the improvements and new features in version 3.1" (Microsoft, 1991). It also saw the introduction of TrueType, a font technology which Microsoft had developed with Apple in an effort to circumvent the licensing fee that both companies had to pay to Adobe for using PostScript. Windows was becoming more international too, with a version supporting the Cyrillic alphabet for Central and Eastern Europe, closely followed by versions for Japan and China. According to Microsoft's Japanese office, nearly one and a half million copies of the Japanese language Windows 3.15 were sold in its first year (Kouyoumdjian, 1994).

Much of this demand was from people wanting to install Windows on their existing machines. However there was also demand from manufacturers who wanted to supply PC compatibles with Windows already installed. Most already had OEM (Original Equipment Manufacturer) agreements with Microsoft that allowed them to purchase MS-DOS at an advantageous price and sell a machine with MS-DOS already installed and configured to load as soon as the customers turned it on. The exact terms of such agreements were confidential, and indeed of little concern to most of the software industry, with the exception of companies such as Digital Research which had their own operating systems to sell. However they were certainly one of the tools that Microsoft used to extend the reach of its products. It was alleged, for example, that PC manufacturers were being offered lower prices if they committed to buying an MS-DOS licence for every PC compatible they sold, regardless of whether it actually had the software installed

(USA v. Microsoft, 1994). In the early 1990s, Microsoft started entering into similar agreements with regard to Windows, with the result that more and more PC compatibles were being sold that powered up straight into the Windows environment.

Windows NT 3.1 was not ready until July 1993 (despite being the first version of this new operating system, it was given the same version number as its 16-bit equivalent). It was not an immediate success as it lacked applications and, at $495, was considerably more expensive than Windows 3.1. However it also came in an Advanced Server version.

The technology needed to network the likes of the PC or the Apple Mac had been around for decades, but most desktop computers still sat in isolation, with users copying files onto floppy disks and walking them to their destination, a solution that quickly became known as a 'sneakernet'. IBM's vision was of offices full of PS/2 machines, all connected through Systems Network Architecture (SNA) to the company mainframe, but that was only of interest to a few large corporations. Rather more useful was Ethernet, a low cost and efficient technology for creating a Local Area Network (LAN) that Bob Metcalfe had developed while working at Xerox PARC in the mid-1970s. Xerox published the Ethernet specification in 1980, in partnership with Digital Equipment Corporation (DEC) and Intel, thereby ensuring its acceptance by the industry. Metcalfe had left Xerox the year before to set up 3Com Corporation, which started selling Ethernet cards for the IBM PC in 1982 (the company name signals its focus on Computers, Communication and Compatibility).

At around the same time, the Utah-based company Novell released a network operating system for PC compatibles called NetWare. This was not an end-user operating system like DOS or Windows, but instead allowed a PC to act as a central server of files and other shared resources which were made available to client computers connected in what became known as a 'client-server' or 'two-tier' configuration. The client machines could be running DOS or Windows; the server, or indeed servers, could be running NetWare. A similar technology became available for the Apple Macintosh a few years later, although unfortunately AppleTalk was not compatible with Ethernet (see page 130).

Another company that took a similar approach started out as Software Development Laboratories (SDL). Founded in 1977 by Larry Ellison, SDL had developed a database management system called Oracle which ran on the PDP-11 and VAX minicomputers. In 1982 Ellison changed the company name to Oracle Corporation, and in 1984 the company released a version that ran under MS-DOS on the IBM PC XT. This was followed in 1986 by Oracle 5.0 which was specifically designed to operate in a client-server configuration. It also came with the Oracle Protected Mode Executive, a small MS-DOS program that allowed the database to operate in protected mode when installed on a machine with an 80286 microprocessor. In the same year Novell released NetWare 286, which was the first operating system to actually run in protected mode.

By the early 1990s, it was quite common to find PCs networked together so they could share files stored on a NetWare server, or share data stored in an Oracle database. It was these products that Microsoft was challenging with Windows NT Advanced Server, which was soon followed by server-side applications such as SQL Server for hosting databases (and so providing an alternative to Oracle), and Exchange Server for managing electronic mail. Windows for Workgroups added local area networking capabilities to Windows 3, allowing Windows users to connect their desktop applications to server applications running under Windows NT. Microsoft could now supply all the software you would need to run a fully connected office.

Meanwhile Digital Research had abandoned its attempt to produce a multi-tasking operating system and, realising that no operating system could succeed in the world of the PC compatible unless it could run DOS applications as effectively as MS-DOS itself, concentrated on producing a direct MS-DOS competitor. Digital Research was supported in this by a number of hardware manufacturers who resented what they perceived as Microsoft's monopoly. There was also a feeling that Microsoft had become complacent and done little to improve MS-DOS since its launch back in 1981.

The result was DR DOS 3.31 which was launched in May 1988, just days after the release of Windows 2.1 (this was actually the first version of DR DOS but it was given the version number 3.31 as that was the current version of MS-DOS at the time). It was followed by DR DOS 5.0 in May 1990, close to the launch date of Windows 3.0.

Both of these did offer advantages over contemporary versions of MS-DOS. DR DOS 5.0, for example, used the High Memory Area available on an 80286 processor to increase the amount of memory available to DOS applications, and could free up even more memory when running on an 80386-based machine. However Digital Research faced a problem as it could never prove that DR DOS was 100 per cent compatible with MS-DOS, and Microsoft was unlikely to endorse it, so users were always faced with the possibility of a DOS application that would not work properly with DR DOS. The problem was compounded by the success of Windows 3.0. Although Windows 3.0 performed many of the tasks expected of an operating system, it was actually a DOS program and still relied on MS-DOS to perform many functions. Indeed aspects of MS-DOS 5.0, which appeared in June 1991, were specifically designed to improve the operation of Windows (Fairhead, 1992, p. 278).

Although such a tight synergy put Digital Research at a disadvantage, Microsoft had a right to ensure that MS-DOS and Windows worked well together. However the issue was highlighted when it was discovered that a 'beta' version of Windows 3.1 (a pre-release version widely distributed for test purposes) generated a 'non-fatal' error message if it detected an operating system other than MS-DOS, or the PC-DOS variant distributed by IBM, and suggested that the user should contact technical support. Furthermore, the code that generated the error was encrypted and deliberately written so as to hinder any attempt to find out what actually triggered the error (Schulman, 1993). In the event, the code was disabled when the product was finally released, and Windows 3.1 proved able to run under DR DOS 5.0. However highlighted the fact that Microsoft was trying to discourage use of DR DOS. As an email from Microsoft senior vice president Brad Silverberg that surfaced later stated, "What the [user] is supposed to do is feel uncomfortable, and when he has bugs, suspect that the problem is DR DOS and then go out to buy MS-DOS" (Goodin, 1999).

Novell purchased Digital Research in 1991 and continued developing DR DOS as Novell DOS until 1994. Then, in July 1996, the Utah-based company Caldera bought the rights to DR DOS from Novell for $400,000 and immediately sued on the basis of "illegal conduct by Microsoft calculated and intended to prevent and destroy competition in the computer software industry." The suit claimed:

> Unless restrained by order of this court, Microsoft will permanently destroy competition in the DOS Market in the microcomputer software industry, and Caldera will be artificially and illegally prevented from realizing the full financial potential of the DOS Business. (Caldera Inc., 1996)

Caldera further claimed that, as a result of Microsoft's anticompetitive practices, sales of DR DOS had fallen from $15.5 million a quarter to just $1.4 million a quarter by the end of its financial year in October 1992. In a familiar story, Microsoft vigorously contested the case but finally settled out of court in January 2000 on payment to Caldera of a substantial but undisclosed sum.

Another company unhappy at the success of Windows was Apple. Apple had not been impressed with Windows 2.0, particularly as it was to form the basis for Presentation Manager, which would be backed by IBM. The company's response, in March 1988, was to file a suit against Microsoft at the federal court in San Jose for breach of copyright, claiming that the original 1985 agreement which the two companies had signed did not extend to various new features that had appeared in Windows 2.03. Particularly at issue was the ability of Windows 2.03 to display windows that overlapped. These had not been a feature of Windows 1.0, and so Apple claimed that they were not covered by the original agreement. The court came down on the side of Microsoft, stating that all but ten of the features that Apple claimed breached copyright were covered by the original agreement, which clearly covered "present and future" versions of Windows, and that the remainder were either not Apple's to copyright in the first place, or were not subject to copyright as they represented the only or obvious way for a particular idea to be implemented.

Apple appealed, and then in December 1989 – presumably worried that Apple might benefit from a settlement that was rightfully theirs – Xerox filed a suit against Apple claiming more than $150 million in damages and royalties owed because of unlawful use of elements from the Xerox Star in the user interfaces of the Apple Lisa and Macintosh. Apple denied any infringement, pointing out that it is the way in which an idea is expressed that is copyrighted, not the idea itself. Indeed Apple's attorney, Jack Brown, went so far as to suggest that the idea that Xerox had originated the Macintosh was "as preposterous as a beaver taking credit for the Hoover Dam" (Pollack, 1990). By the following March, Xerox's claims had been dismissed. Apple's suit against Microsoft dragged on until 1994, when it too was dismissed, and Apple's right to further appeal denied.

Looking back, Bill Gates described the early conflict between OS/2 and Windows as the "Battle of the Weaklings" because, until the release of Windows 3.0, neither had any particular relevance to the ordinary PC user (Gates B. , Bill Gates Interview, 1993). However just a few years later it was clear that the future lay with Microsoft. Apple did fight back, taking out an advert to coincide with the launch of Windows 3.0 that suggested, "Now that everyone agrees how a computer should work, try the only one that actually works that way." The advert continued:

> A Macintosh is a Macintosh from the inside out. Conceived from the chip up to work intuitively and visually ... Because we engineer both the hardware and its operating software, Macintosh runs with the smooth speed and precision you'd expect from any perfectly integrated design. And because Macintosh isn't a "graphical" shell grafted on top of a character-based system, it doesn't expend lots of expensive computing power trying to do something it wasn't designed for. (Apple Computer, 1990)

Apple had a point, and indeed few would have disagreed that the Macintosh was a more elegant solution, but at the time an Apple Macintosh II with just a 40MB hard disk and a colour screen cost at least £3,500 and could not run the many DOS applications on which the business world had come to rely. By 1994 it was estimated that over 90 per cent of desktop computers were running a version of either MS-DOS or Windows, and connecting to an increasing number of back-room servers that were running a version of Windows NT. However the world was about to change again.

12.   Owning the Web

As microcomputers made their way into people's homes through the late 1970s and early 1980s, they weren't only being used to play games and write programs. Enthusiasts were also connecting them to dial-up modems and exploring the outside world, where they were discovering all sorts of weird and wonderful things.

A dial-up modem works by converting digital information into audible sounds, so enabling one computer to communicate with another across an ordinary telephone line. Anyone with a home computer, a dial-up modem and suitable software could connect to any other computer that was connected to the telephone system, provided they knew its phone number. However this was very different to the world of broadband and the World Wide Web that we are familiar with today. At the beginning of the 1980s, such modems could just about manage to transfer 300 bits per second, which meant that it took at least 30 seconds to fill a screen with text. Modems capable of 1,200 bits per second followed, but it was still a laborious process.

Furthermore, the landscape into which they connected was more like the Wild West than the bustling online metropolis we inhabit today. Each computer system that you might connect to had its own way of operating and, even if it did welcome guests, it was frequently offline or engaged. Nevertheless that was all part of the fun – you never knew what you would find next or who you would come across on your travels, and it was always exciting, even if it did require a great deal of patience.

Particularly popular were bulletin boards where people could leave messages and carry out conversations with like-minded people on a wide range of topics. These had come a long way since Lee Felsenstein and Efrem Lipkin had set up Community Memory in San Francisco back in 1972, but at the forefront were the amateur systems that were being put together in people's homes using little more than a microcomputer and a modem. CBBS (Computerised Bulletin Board System), which was run from Chicago by Ward Christensen and Randy Suess and went live in February 1978, is generally recognised as the first.

The idea came to Christensen while snowed in at home during the Great

Blizzard of 1978 and not able to get to his office. What Christensen envisaged was a program that would effectively turn his microcomputer into an answering machine, so that members of the Chicago Area Computer Hobbyist's Exchange (CACHE), to which they both belonged, could deposit articles for the group's newsletter. From this came the idea of an electronic bulletin board which would allow anyone to dial in and read messages or contribute to on-going discussions.

Suess was responsible for putting together the hardware, which was based around an IMSAI 8080 processor board with just 24K of memory, a floppy disk drive and a modem that could communicate at either 110 or 300 bits per second. Christensen completed the program code in just two weeks, and soon CACHE members were leaving messages and starting discussions. Users were asked for a first and last name, but there was no need to enter a password unless you wanted to identify yourself as a system operator (or 'sysop'), which gave you access to various maintenance functions. Full details were published in the November 1978 issue of *Byte* magazine, together with the bulletin board's phone number.

Calls outside your immediate area code were very expensive at the time, and CBBS' single modem meant that the phone number was frequently engaged. As Christensen wrote in the *Byte* article:

> We would like to see other experimenters or clubs implement such a system. These bulletin boards could then become *nodes* in a communication network of automated message and program switching.

Such a network would require something more sophisticated, but Christensen got the ball rolling by selling his software for $50 to anyone interested in setting up on their own (Christensen & Suess, 1989). Soon bulletin boards were listing the phone numbers of hundreds of BBSs across the United States, and boards were being set up in other countries as well: the July 1984 issue of *Your Spectrum* listed 16 amateur bulletin boards in the UK with names like CBBS Cumbria, Forum-80 Hull and TBBS London (the first part of the name indicating the software being used to run the board).

New software was appearing, too. In 1981, Jeff Prothero (known online as 'Cynbe ru Taren') wrote a BBS program called Citadel which introduced the concept of 'rooms'. New users found themselves in the LOBBY, from which they could discover other rooms or create their own with a name appropriate to the topic being discussed. Such bulletin boards were social places where people could interact unhindered by looks or gender. As one young enthusiast put it:

> It's just the idea! Like I'm talking to people and they could be anywhere and
> I don't know who they are! That's really neat! Even if a lot of them turned out to
> be really boring. (Scott J. , 2005)

Initially, each board operated in isolation, serving a community that largely lived within the same area code as the bulletin board itself. However in 1984, Tom Jennings rewrote the Fido program that he was using to run his BBS in San Francisco so that one installation could exchange messages with another, creating a network that he called FidoNet. Most of the time, each node operated as a standalone bulletin board, taking in messages from local users. However, at a designated hour each day, each node would exchange messages with its immediate neighbours, so effectively creating a single BBS that spanned the country and eventually the world. Within six months, FidoNet contained 150 nodes, and by 1986 it spanned nearly a thousand. However it was very slow, with messages taking days or even weeks to traverse the system. As Jennings himself describes:

> You dial a number and you enter your name and password. It says hello, and
> then you can go to the messages and you can read the messages. And then you
> could add one. And if you waited long enough – and I had to say, 'months' –
> other people would have called in and left messages, and after a few months, you
> have a conversation ... It is stupid - it's unbelievably stupid! (Scott J. , 2005)

The problem with FidoNet, and indeed with any communication system that involves dial-up modems, is that it operates across conventional telephone lines, and that means establishing and maintaining a dedicated connection for the exclusive use of the two parties involved. This makes sense if it's two people holding a conversation, but it is an extremely inefficient way of exchanging bits and bytes between large numbers of computers. Far more efficient is a technique known as 'packet switching' which was developed in the early 1960s and first adopted in earnest by ARPANET, the network created by the Advanced Research Projects Agency (ARPA) to link the computer networks of organisations involved in the Cold War and the Space Race.

A packet switching system has analogies with a postal system, although the analogy does not bear too much scrutiny. In a communications system that uses packet switching, the data that makes up a message is divided into 'packets' or 'datagrams' of a particular length, perhaps a few kilobytes at a time, and each packet is attached to a 'header'. The header effectively acts as an envelope, in that it contains the addresses of the sender and the receiver. It also details the length of the packet, the number of packets that make up the complete message and the position of this particular packet in the sequence. If this was a postal system, you could then drop the envelopes into your local post box, or indeed split them up and drop them into a number of different post boxes, confident that the postal system will ensure that they all eventually reach the intended receiver, and that the receiver will be able to reconstruct the original message once they all arrive.

The advantage of such a system is that it is no longer necessary to establish a dedicated communication path and hold it open until all the data has been successfully transferred. Returning to our postal analogy, thousands of letters can be transported by a single vehicle between one sorting office and another, even though the final destination of each one may be different. All that is necessary is that each node on the network reads the destination address of each packet as it arrives, and send it on its way. This all makes for a network that is far better suited to putting large numbers of computers in touch with each other.

By 1981, ARPANET had established connections to more than 200 mainframes and minicomputers at major universities. These included the Stanford Research Institute, where Douglas Engelbart had developed his NLS system, the University of Utah, where Ivan Sutherland and Alan Kay had worked, Massachusetts Institute of Technology and Harvard University, as well as various defence organisations such as the ARPA offices at the Pentagon, National Aeronautics and Space Administration (NASA) and North American Aerospace Defense Command (NORAD). The network also extended across the Pacific to bring in ALOHAnet at the University of Hawaii, and across the Atlantic to University College London via the Norwegian Seismic Array (NORSAR), which monitored Soviet nuclear tests.

The computers at these institutions were intended for use by authorised personnel only. However, as microcomputers became more common, universities in particular started to provide dial-up access so that staff could work from home. There were also public networks such as Packet Switch Stream from British Telecom and Telenet in the United States, which offered dial-up access to subscribers. Armed with a valid user identity and password, you could dial in to a computer on one of these networks and start exploring. Indeed with a suitable set of user identities and passwords, and a good understanding of the various systems that you might encounter, you could move through the network and on to computers connected to other networks in other institutions.

This was, of course, risky unless you had explicit permission to do so. However for many the temptation was too great and the law too vague. As Kevin Mitnick, who was arrested in 1988 for breaking into corporate networks and later went on the run from the FBI having broken the terms of his probation, put it:

> The thrill was being somewhere you shouldn't be, and trying to remain undetectable. It's about forbidden knowledge; about pranksterism; about trying to outsmart the other. (Lee, 2001)

A typical story is that of Ron Austin and Kevin Poulsen (who styled himself 'Dark Dante'). In 1982, these two teenagers from Santa Monica managed to guess the password for an ARPANET user named UCB (as in University of California, Berkeley). With the help of a TRS-80, a Commodore VIC-20 and an instruction manual for the popular Unix operating system, they used this information to access computers at the National Research Laboratory in Washington, NORSAR in Norway, NORAD and RAND Corporation – a 'think tank' organisation used by the US military which became famous for developing the strategy of nuclear deterrence known as Mutually Assured Destruction (MAD).

Eventually, noticing an uncharacteristic increase in activity on the UCB account, the university alerted the FBI who managed to track down Austin. Just hours before his arrest, he is supposed to have signed off from the online chat session he was having with Poulsen with the line, "Got to go now, the FBI is knocking at the door." Charged with 'malicious access', Austin served two months in prison. Poulsen's TRS-80 was confiscated, but he was under 18 which meant that he was too young to be prosecuted. Shortly afterwards, aware that one of the systems Poulsen had hacked was its own, Stanford Research Institute offered him a job as a security consultant with a salary of $35,000 a year (Littman, 1993).

Such exploits were celebrated in the film *WarGames*, which was released in 1983, shortly before Poulsen and Austin were arrested and gaining them the headline 'UCLA Wargames Arrest' in the *Los Angeles Herald Examiner*. The hero of the film is a teenager, played by Matthew Broderick, who uses his IMSAI 8080 to impress a classmate (Ally Sheedy) by breaking into the school's computer and changing their grades. They go on to hack into what they think is a computer games company but is actually a NORAD missile-command computer called WOPR (as in War Operation Plan Response) and start playing what appears to be a game called Global Thermonuclear War but is actually the real thing. They are arrested as suspected Russian agents but eventually manage to convince the authorities to take them seriously and, in the nick of time, persuade WOPR that Global Thermonuclear War is a game that can only be won by not playing it in the first place.

The film became an enduring success, partly because it succinctly captured the prejudices and fears prevalent at the time. This was before the widespread computer virus attacks and malicious identity theft that we have today, so hacking was widely viewed as a teenage prank – irresponsible and misguided, but not malevolent. The Broderick character doesn't want to hurt anybody, he just wants to impress Ally Sheedy; and indeed the evidence suggests that hackers like Poulsen and Austin weren't particularly interested in the documents they came across as they roved around ARPANET – they just wanted to demonstrate that they could do it. The film also highlights the tensions of the Cold War. This was the year in which Ronald Reagan labelled the Soviet Union an "evil empire" and many were fearful of a military over-reliant on technology. As Mitnick later stated:

> That movie had a significant effect on my treatment by the federal government. I was held in solitary confinement for nearly a year because a prosecutor told a judge that if I got near a phone, I could dial up NORAD and launch a nuclear missile ... I think the movie convinced people that this stuff was real. (Brown S. , 2008)

Although most of the activities that a hacker might get up to were covered perfectly adequately by existing legislation against trespass, fraud and theft, new laws were quickly put in place. The United States Congress passed the Computer Fraud and Abuse Act in 1986 and the United Kingdom followed with the Computer Misuse Act in 1990, which made it illegal to try to gain unauthorised access to a computer, even if the attempt is unsuccessful.

As more people came online, large-scale communities such as CompuServe and the WELL (Whole Earth eLectronic Link), or CIX (Compulink Information eXchange) in the UK, started to flourish. These were dial-up services, each offering email and a range of discussion forums to subscribers through an increasing numbers of dedicated phone numbers across America, and later in Europe. CompuServe was the largest, despite subscription charges that earned it the nickname Compu$erve or CompuSpend (Raymond, 1991). In 1984 it introduced the Electronic Mall with the tag line, "Let your fingers do the shopping". Shoppers could buy from an increasing number of suppliers by entering areas within the Mall with names like the Hardware Store, the Software Boutique, the Record Emporium or the Gardening Shed (CompuServe, 1984).

Many of these services also linked to Usenet, a distributed conferencing system that worked much like FidoNet, but a great deal more efficiently as it could run across packet switched networks, as well as the conventional telephone system. Usenet started life in late 1979 as a means of exchanging information between the academic community of North Carolina. Discussions took place in special interest forums or 'newsgroups' which were organised in a broad hierarchy: groups with names like 'net.general' or 'net.test' were intended for general discussions, while those with names like 'fa.human-nets' or 'fa.network-hackers' were copied from ARPANET message lists (hence 'fa'). This soon got out of control as the number of groups exploded, including many of dubious academic merit such as 'net.jokes' or 'fa.wine-tasters'. As a result a new 'mod' category was created to indicate newsgroups in which the content was controlled by someone acting as a moderator. Nevertheless the number of messages continued to escalate to the extent that it was threatening to overwhelm many of the systems involved in forwarding the message lists.

One way in which the traffic could be controlled was for system managers to forward only a selection of newsgroups, but this inevitably proved controversial, as it amounted to censorship. The result was what became known as the Great Renaming of 1986 in which a new hierarchy was hammered out by a group of system managers nicknamed the Backbone Cabal in often acrimonious discussions. Seven top-level hierarchies were eventually agreed, including 'comp', 'misc', 'news', 'rec', 'sci', 'soc' and 'talk' (for more controversial subjects). But even that wasn't enough, and the following year saw the Breaking of the Backbone Cabal following its refusal to forward the newsgroups 'rec.sex' and 'rec.drugs'. In May 1987, while dining at G.T.'s Sunset Barbecue in Mountain View, California, John Gilmore and Brian Reid decided to create a newsgroup called 'alt.drugs' without going through the procedures that had become accepted practice, and propagate it through the system in a way that avoided the Cabal (i.e. using an 'alternative' route). The following year, Reid sent the following email:

> To end the suspense, I have just created alt.sex. That meant that the alt network now carried alt.sex and alt.drugs. It was therefore artistically necessary to create alt.rock-n-roll, which I have also done. I have no idea what sort of traffic it

will carry. If the bizzarroids take it over I will rmgroup it or moderate it; otherwise I will let it be. Brian Reid, 5th thoracic (Reid, 1993)

The 5th thoracic is a vertebra on the human backbone, reminding the email's recipients that Reid, too, was a member of the Backbone Cabal.

Usenet and the 'alt' newsgroups in particular blossomed, with groups such as 'alt.startrek.borg', 'alt.sex.bondage', 'alt.god-talk' and 'alt.tasteless'. A popular retreat was 'alt.callahans' in which you were expected to behave as though you had just stepped into a bar similar to that described in the *Callahan's Place* series of books by science fiction writer, Spider Robinson. It was regarded as a friendly relaxed environment where you could share your troubles and joys with friends. Hugs and backrubs were exchanged, and the traditional weapon employed against 'inveterate punsters' was the peanut. If you wanted to make a speech, you used a Soapbox, or you could wander through the French Doors to other settings, such as Didi's Meadow, which was described as "the perfect place for a picnic" (Danger Mouse, 1995).

By the early 1990s, Usenet could offer over 6,000 newsgroups and was being used by several million people across the world, prompting one commentator to remark:

> Usenet is like a herd of performing elephants with diarrhoea – massive, difficult to redirect, awe-inspiring, entertaining, and a source of mind-boggling amounts of excrement when you least expect it. (Spafford, 2009)

And then, of course, there were the online games, perhaps the best known being MUD (Multi-User Dungeon). This was a text-based adventure game written by Roy Trubshaw to run on a DEC minicomputer at Essex University in 1978, and augmented by Richard Bartle. It was originally accessed only by fellow students, but the system administrators allowed outsiders to log in at night through JANET (Joint Academic Network), a packet switched network that went live in 1983, connecting some 50 academic institutions across the UK. MUD became extremely popular as players turned themselves into fantasy characters, exploring the lovingly-described landscapes and helping or hindering each other as they strived to achieve the coveted rank of Wizard, or even Arch-Wizard.

By the mid-1980s, 'inter-networks' such as ARPANET and JANET were linking an increasing number of computer networks together using various forms of packet switching technology, and facilitating the spread of online applications such as email and Usenet. This nascent Internet was far from homogenous and still ostensibly limited to academic use. However microcomputers and modems were spreading through people's homes, and the various government and academic bodies involved in its development were beginning to realise that they had a tiger by the tail.

The first step towards what we have today was the insistence by the US Department of Defense that ARPANET should move from its existing system for handling packet switching, which was based around a piece of software known as the Network Control Program (NCP), to a more flexible system known as TCP/IP which would be far better suited to large-scale networks. TCP/IP defines two protocols for dealing with the data that is to be sent across a network. One is the Transmission Control Protocol (TCP), which describes the way in which the data is divided into packets and how it is reassembled at the other end. This is partnered by the Internet Protocol (IP), which defines the structure of the 'envelope' and the way in which the source and destination addresses are specified. It is this protocol which is primarily concerned with the way in which computers from one network connect to those in another, and as IP is almost always used with TCP, the combination is also frequently and confusingly referred to as the Internet Protocol.

The US Department of Defense decreed that all systems connected to the ARPANET should be using TCP/IP by 1 January 1983, and in the event the target was missed by only a few months – partly thanks to TCP/IP being incorporated into Unix, the operating system used by many academic minicomputers (Postel, 1981). Shortly after the changeover was completed, around half of the nodes within ARPANET were moved to a separate network called MILNET, with just a few carefully controlled gateways linking the two. Non-classified military data was handled by MILNET, leaving 45 academic and scientific institutions connected to ARPANET (Computer History Museum, 1997).

Then in 1984 the National Science Foundation, which at the time had an annual budget of over $1 billion to spend in supporting education and non-medical research across the United States, initiated the construction of five supercomputer centres. These were to be linked by NSFNET, a high-speed network designed around TCP/IP so that it could connect not only to ARPANET but also to other TCP/IP networks that might be developed elsewhere. In 1987, the NSF chose to contract out the work of upgrading the system to a consortium that included IBM and MCI (Microwave Communications Incorporated), marking the start of an effort to involve commercial companies. The result was a collection of interoperating networks that expanded rapidly and was increasingly being referred to as the Internet.

Connections to Canada and the Scandinavian countries were established quite quickly, but other countries were slower to link up, particularly those that had already made big investments in networks based on protocols other than TCP/IP. Adoption was also hindered by a long-standing effort to establish a networking standard called Open Systems Interconnection (OSI), and indeed OSI had been published by the International Organisation for Standardisation (ISO) in 1984. Many were keen to adopt it, particularly in Europe, but others argued that it was too complicated. In any case, the pragmatic nature of TCP/IP, and the huge investments being put in to TCP/IP networks in the United States, was rendering OSI irrelevant.

JANET was doing a very good job of linking British academic institutions, but was based on what were known as the Coloured Book protocols, thanks to the different coloured covers used to identify the books that described each of its eight protocols. JANET did have a link to the Internet through the ARPANET connection at University College London, but this was very limited so in 1991, once efforts to implement OSI had been abandoned, the JANET IP Service was introduced and JANET effectively became part of the Internet.

Another important European network was that used by the European Organisation for Nuclear Research. This organisation is best known for the laboratory outside Geneva which now houses the Large Hadron Collider, although CERN (as it is more commonly known) actually includes laboratories and institutions spread right across Europe. The networks linking these various organisations had been pretty chaotic, with much argument between different factions, but by the mid-1980s it was becoming obvious that TCP/IP offered the best solution for integrating the various systems that were in use (Segal, 1995). By 1989, adoption of TCP/IP was sufficient for CERN to open connections to the Internet; and in the following year, as part of its European Academic Supercomputing Initiative (EASI), IBM sponsored a high-speed link between the NSFNET node at Cornell University, New York State and the CERN laboratory in Geneva, establishing CERN as one of the main Internet gateways connecting the two continents (Streibelt, 1992).

By 1990, when ARPANET was finally decommissioned, the Internet spanned nearly 300,000 computers within some 2,000 organisations across more than 30 countries (Zakon, 2011). NSFNET remained crucial to its functioning but, partly in an effort to prevent it from becoming overloaded, and partly to justify its public funding, use of the NSFNET backbone had always been restricted to the needs of those who were directly involved either in education or in research: "extensive use for private or personal business" was deemed unacceptable. As the number of people accessing the Internet increased – particularly the burgeoning hordes of hobbyists with their microcomputers and modems – such a policy was clearly becoming unrealistic.

Having ensured that TCP/IP was firmly established, to the extent that the Internet could now be defined as the collection of networks that were based on and linked together using TCP/IP, the NSF was in fact keen to hand over responsibility to private companies such as IBM and MCI, and in 1988 initiated a series of conferences under the title 'The Commercialization and Privatization of the Internet' (Internet Society, 2012). Then, in 1992, it received authorisation to "support the development and use of computer networks which may carry a substantial volume of traffic that does not conform to the current acceptable use policy" (Cringely, 1998).

In the same year, Cliff Stanford launched Demon Internet, the first Internet Service Provider (ISP) in the UK to offer dial-up access to the Internet for just "a tenner a month" to anyone who wanted it. Demon Systems, and the many other ISPs that soon followed suit, were taking liberties, but as Stanford put it:

> ... it was never a real problem. The people trying to enforce [the restrictions] weren't working very hard to make it happen, and the people working to do the opposite were working much harder.

Such services became popular, but as he recalls:

> The question we always got was: "OK, I'm connected – what do I do now?" It was one of the most common questions on our support line. We would answer with "Well, what do you want to do? Do you want to send an email?" "Well, I don't know anyone with an email address." People got connected, but they didn't know what was meant to happen next. (Burkeman, 2009, p. 11)

That was all about to change, thanks to a project that Tim Berners-Lee, a British contractor working at the CERN laboratory in Geneva, was tinkering with in his spare time.

Hypertext was by no means a new idea. Ted Nelson coined the word in 1963, although what he described was rather more ambitious than what we have now. Douglas Engelbart demonstrated hypertext in 1968 during his famous 'Mother of all Demos' in which he showed how a word within one document could become a link to another, so that clicking on the word with a mouse would cause the linked document to open. It was an extremely simple idea, and indeed Ted Nelson later described it as "a discovery, not an invention, because it's obvious" (Cringely, 1998).

One of the more surprising things in the history of computing is that no one before Berners-Lee seems to have realised just how well suited the concept was to the rapidly expanding Internet. Instead, hypertext was seen as a niche technology used in documentation systems and the odd interactive novel. In 1987, Apple had gone so far as to bundle a free copy of a product called HyperCard with every Macintosh. Created by Bill Atkinson a couple of years earlier, HyperCard allowed you to create a stack of virtual cards which could be linked to each other through underlined words or 'hotspots', much like modern web pages. It was great fun, but it didn't occur to anyone that such links could reach across a network. As Atkinson later admitted:

> I missed the mark with HyperCard. I grew up in a box-centric culture at Apple. If I'd grown up in a network-centric culture ... HyperCard might have been the first web browser. My blind spot at Apple prevented me from making HyperCard the first web browser. (Lasar, 2012)

While working at CERN during the mid-1980s, Berners-Lee became concerned at the haphazard way in which technical papers, instruction manuals and other important documentation were scattered across the largely incompatible computer systems used by the various research teams. Various solutions had been suggested for organising this information into something more coherent, but all had failed because researchers had more important things on their minds. As Berners-Lee quickly realised:

> I would have to create a system with common rules that would be acceptable to everyone. This meant as close as possible to no rules at all. (Berners-Lee, 1999, p. 17)

Getting CERN interested proved difficult. Only in 1990, after having two proposals ignored, was he finally given the funds to purchase a desktop computer that would allow him to try out his new ideas. Having got the machine (a NeXT computer from the company that Steve Jobs had set up after leaving Apple), Berners-Lee set about creating the first web browser, which he called WorldWideWeb; and the first website, which he named 'info.cern.ch'. Both were actually running on the one machine, but they were connected by protocols that could span the world.

Berners-Lee realised that his protocols would only be accepted if they were extremely simple. The most important was the Universal Resource Identifier (URI), which is a string of text that uniquely identifies a particular resource, such as a program or a document. For example, the location of the very first web page ever created was defined as 'http://info.cern.ch/hypertext/WWW/TheProject.html'. This describes a file named 'TheProject.html' which was originally to be found within the '/hypertext/WWW' subdirectory on his machine, and served as what we would now call the 'home page' to the documentation surrounding his project.

The first four characters of this URI reference the Hypertext Transfer Protocol (HTTP) which defines how a browser requests a document from a website, and how the site should respond. The request simply consisted of the word 'GET', followed by the URI of the document concerned. The response would be the document itself. In addition Berners-Lee devised a system for 'marking up' a conventional text file which he called Hypertext Markup Language (HTML). This defines a set of 'tags' that can be inserted into the text to indicate formatting (such as bold or italic), or to create links between one document and another. The links themselves are defined by their URIs.

Particularly important to Berners-Lee's vision was that the machine running the web browser did not need to be compatible with the machine serving up the website. The web server simply had to know how to respond to the GET command, while the web browser simply had to understand HTML. The idea was straightforward but, as Berners-Lee explains:

> What was often difficult for people to understand about the design was that there was nothing else beyond URIs, HTTP and HTML. There was no central computer 'controlling' the Web, no single network on which these protocols worked, not even an organisation anywhere that 'ran' the Web. The Web was not a physical 'thing' that existed in a certain 'place'. It was a 'space' in which information could exist.
>
> I told people that the Web was like a market economy. In a market economy, anybody can trade with anybody, and they don't have to go to a market square to do it. What they do need, however, are a few practices everyone

has to agree to, such as the currency used for trade, and the rules of fair trading. The equivalent of rules for fair trading, on the Web, are the rules about what a URI means as an address, and the language the computers use ... (Berners-Lee, 1999, p. 39)

Berners-Lee's colleague Robert Cailliau had also bought a NeXT machine, and on Christmas Day 1990, he managed to create a connection across the Internet to the info.cern.ch website running on Berners-Lee's machine.

It took some time for interest to grow, partly because Berner-Lee's WorldWideWeb program only ran on the relatively obscure NeXT machine, but over the next few years, various people within the academic community wrote and distributed web browsers that could run on a variety of mini and mainframe computers. Berners-Lee made the code for his web server publicly available, and new websites started to appear outside the CERN community. One of the most impressive at the time was created by Frans van Hoesel to show off paintings lent to the Library of Congress by the Vatican, allowing you to wander through various 'rooms' and view the paintings on screen (van Hoesel, 1994).

By the start of 1993, around 50 web servers were up and running, and Berners-Lee had made some progress in establishing his all-important protocols as officially accepted standards. However the Web was growing in a rather different fashion to the way he had envisaged. His original WorldWideWeb program didn't just allow you to view web pages; you could also edit and add your own. For Berners-Lee, the Web was somewhere people could collaborate, as well as browse. However others were content to release their software as soon as it was capable of browsing, without bothering to add editing facilities.

The most significant browser to appear during this period was Mosaic, which had been developed by Marc Andreessen and Eric Bina at the National Center for Supercomputing Applications (NCSA). Located within the University of Illinois, where Andreessen was an undergraduate, the NCSA was one of the five supercomputer centres created by the NSF back in 1984 and was directly connected to NSFNET, giving it unprecedented high-speed access to the Internet.

The original version of Mosaic was released in January 1993 and ran under Unix, with versions for Microsoft Windows and the Apple Macintosh following in August. It was important for a number of reasons. As Berners-Lee himself put it:

Marc and Eric did a number of very important things. They made a browser which was easy to install and use. They were the first one to get inline images working – [up] to that point browsers had had varieties of fonts and colors, but pictures were displayed in separate windows. This made web pages much sexier ... Mosaic was the easiest step onto the Web for a beginner, and so was a critical element of the Web explosion. (Berners-Lee, Frequently Asked Questions)

And it was indeed an explosion. The NCSA distributed Mosaic free of charge, and by the end of the year, following a prominent article in *The New York Times*, more than a thousand copies were being downloaded every day. The number of web servers was also increasing, to 500 by October and 1,500 the following June, by which time it was claimed that at least a million copies of Mosaic had been downloaded (Wolf, 1994). The content available from these servers was not all technical or academic, but included websites set up by the likes of Hyatt Hotels, Tupperware and Volvo, as the importance of this new medium became apparent to a wider audience.

Mosaic was free if used for private, academic or research purposes. However the NCSA was also happy to license the source code on a non-exclusive basis, allowing software companies to enhance and sell versions under their own name. Such licences cost $100,000 plus $5 for each copy sold, and customers included Fujitsu, which announced a Japanese version in June, and Spry, which included a version of Mosaic as part of its Internet in a Box (IBox). At $149, this included all the software that you needed to link up to a dial-up Internet service and start browsing the Web.

Unhappy that others were profiting from a project into which he had put so much effort, and for which he was getting so little credit (he hadn't even been mentioned in *The New York Times* article), Andreessen left the NCSA at the end of 1993. However a few months later he was contacted by Jim Clark, who had made his money from high-end graphics systems, and the two of them set up Mosaic Communications.

As he had written much of Mosaic in the first place, Andreessen did not feel it necessary to pay a licence fee to the NCSA in order to produce something that "looks similar and works better" (Wolf, 1994). By May 1994, most of the remaining staff who had worked on Mosaic had also left the NCSA to join Andreessen's new company, and the NCSA was decidedly unhappy. Not wishing to fight a lawsuit, Clark and Andreessen decided "to accommodate concerns expressed by the University of Illinois" (Netscape Communications, 1994), and changed the company's name to Netscape Communications. The new browser became Netscape

Navigator, while the original was generally referred to as NCSA Mosaic to distinguish it from other licensed versions.

Netscape's plans were ambitious. Andreessen saw Mosaic becoming not only the user interface to the Web but also to the desktop itself, doing away with Microsoft Windows and Mac OS to give users direct access to applications running on remote machines across the Internet. As he told *Wired* magazine, "One way or another, I think that Mosaic is going to be on every computer in the world" (Wolf, 1994). The details were vague, but Netscape wasn't worried: one million people may have already downloaded Mosaic but there were an estimated 25 million people connected to the Internet, so there was plenty of scope left to make money. As Jim Clark put it:

> By the time we had our product on the marketplace it would be 50 million people: you've got to be able to make money with 50 million people using your product. That was the sum total of the business plan. (Cringely, 1998)

Like NCSA Mosaic, Netscape Navigator was free of charge for personal, academic or research use, provided you were prepared to download it over the Internet. Commercial users could purchase a copy for $99, with volume discounts available to companies (Mosaic Communications, 1994). Alternatively you could buy a boxed retail copy, ready to install, for $49.

Although encouraged by the growth that their activities had inspired, Berners-Lee was concerned that the approach taken by Andreessen and the NCSA was moving the focus away from the Web to what was, after all, just one solution to the browsing problem, to the extent that people were actually talking about Mosaic as though it was the Web itself. Following consultation with many people in the industry, he came to the decision that the best route would be to found an independent consortium. He announced his intention at the first World Wide Web conference, a three-day affair held in May 1994 at the CERN laboratories in Switzerland which was attended by 350 people and dubbed by the press "the Woodstock of the Web". Following the agreement of CERN, which owned the work he had done, the World Wide Web Consortium (W3C) was established at MIT that October with Berners-Lee as director. As he explains:

> My motivation was to make sure that the Web became what I'd originally intended it to be – a universal medium for sharing information. Starting a company would not have done much to further this goal, and it would have risked the prompting of competition, which could have turned the Web into a

bunch of proprietary products ...

... by following the consortium route ... I'd be free to really think about what was best for the world, as opposed to what would be best for one commercial interest. (Berners-Lee, 1999, p. 91)

The Consortium had no legal powers, but early members included Netscape, IBM and Hewlett-Packard, and anyone involved in the Web was encouraged to join. As a result, W3C Recommendations do still effectively define the standards by which the Web operate.

Despite the growth of the Internet, in the early 1990s the big money was still being made by online services which operated in isolation as a series of 'walled gardens', offering content to primarily dial-up customers who made direct connections through the telephone system. The oldest was CompuServe, which charged its members by the hour. In 1992 it introduced a proprietary browser called CompuServe Navigator that provided a more colourful and friendly user interface, and by 1993 claimed more than 1.5 million subscribers. America Online (AOL) charged a flat monthly fee and by 1993 could boast over half a million subscribers who could access its services through a friendly graphical user interface available for both Windows and the Apple Macintosh. The other big player was Prodigy, a joint venture between Sears and IBM. Launched in 1986, it had attracted half a million users by 1990. Again it was accessed through Prodigy's own software, which cost $49.95, and users were charged $9.95 a month for an annual subscription.

Now that Windows was to be found pre-installed on most PCs, Microsoft realised that it was in a unique position to cash in on this market. It therefore started work on Microsoft Network (MSN), an online service that would offer services similar to CompuServe or AOL but with the advantage –at least in Microsoft's eyes – that users could be invited to sign up as soon as they turned on their new PCs, which would come ready-installed with the necessary software as part of the forthcoming Windows 95.

Microsoft was also becoming aware of the challenge that Netscape posed. In March 1995 it announced a "design environment for online applications" under the codename 'Blackbird'. The press release continued:

Using 'Blackbird', content providers can implement online publications, consumer-oriented applications, business-to-business services, interactive advertising, electronic-commerce applications, and even interactive games for The Microsoft Network. (Microsoft, 1995)

Microsoft went on to make it clear that Blackbird, and in turn MSN itself, would be "more sophisticated" than anything Netscape, or indeed the World Wide Web, could offer. That August, Microsoft launched Windows 95 alongside its MSN service. As Bob Metcalfe of *InfoWorld* commented:

> Expect fierce competition between Microsoft and Netscape. The important question is not whether Microsoft or Netscape wins, but whether we'll end up choosing our Web standards from those developed in an open computing consortium or those controlled by Microsoft. (Metcalfe, 1995)

Microsoft had set itself up to challenge the rest of the industry, but at the same time it was coming to realise that the imminent battle could only harm the company. Gates himself signalled a change of heart in an internal memo entitled 'The Internet Tidal Wave' which was sent to his executive staff in May, just a couple of months after the Blackbird announcement. In it he stated:

> ... I want to make clear that our focus on the Internet is crucial to every part of our business. The Internet is the most important single development to come along since the IBM PC was introduced in 1981. It is even more important than the arrival of the graphical user interface (GUI). (Gan, 2011)

He also accepted that such a stance put MSN in an awkward position:

> ... every on-line service has to simply be a place on the Internet with extra value added. MSN is not competing with the Internet although we will have to explain to content publishers and users why they should use MSN instead of just setting up their own web server. We don't have a clear enough answer to this question today. For users who connect to the Internet some way other than paying us for the connection we will have to make MSN very, very inexpensive - perhaps free. The amount of free information available today on the Internet is quite amazing. (Gan, 2011)

Such a change in direction could not have come at a more awkward time. Microsoft had already made a big fuss about MSN and Blackbird, and Windows 95 was only a few months from launch. Nevertheless, to the confusion and consternation of its many business partners, Blackbird was hastily repackaged as Internet Studio, a set of tools for creating and delivering websites, and MSN was repositioned to become in effect just another website.

More significantly, Microsoft entered into an agreement with Spyglass. The NCSA had transferred to Spyglass all commercial rights to Mosaic the year before, and Spyglass had gone on to create its own Enhanced Mosaic Word Wide Web browser, including a version that ran under Windows 3.1. Under the agreement, Microsoft turned this into Internet Explorer on the understanding that it would pay Spyglass a basic fee plus a royalty on each copy sold. However, although bundled with a collection of add-ons called Microsoft Plus! for Windows 95 which did cost $49.95, Internet Explorer was effectively given away, as it could be downloaded free of charge from America Online, CompuServe and from Microsoft's new website. It was also included in the OEM version of Windows 95 that was distributed to hardware manufacturers, which meant that it was prominently displayed on the desktop as soon as most users booted up their new PCs.

Microsoft attempted to explain and justify its change of heart on Internet Strategy Day, an event for analysts, journalists and its business partners that was held in December 1995. Here, Gates announced his company's intentions to "embrace" existing Internet protocols in all its products, and (more ominously) to "extend" such protocols where it could to offer a richer experience. This change in focus affected almost all of Microsoft's product range, from Windows itself through to its applications and on to other projects such as MSN, with Windows NT on hand to deliver the websites themselves through a new server application called Internet Information Server (IIS). It was a bold move. As *BusinessWeek* put it:

> … in just six months, Gates has done what few executives have dared. He has taken a thriving, $8 billion, 20,000-employee company and done a massive about-face. (Rebello, 1996)

It was also somewhat ironic in that Microsoft's about-face bore comparison to that of IBM 15 years previously, when it had belatedly recognised the threat posed by the microcomputer and hastily assembled the IBM PC. Then it was IBM that represented the establishment and Microsoft was one of the renegade start-ups backing the revolution. Now Microsoft was the establishment, and in danger of being outsmarted by a new band of revolutionaries.

By the mid-1990s, Microsoft had a product range that catered for both sides of a network. However its strategy was to make the most of the computing power available on the desktop, interacting with the server only when it became necessary to access shared data, or to communicate with someone else. This made sense while networks remained slow, but it occurred to Andreessen that, as networks became more efficient, computing power need no longer reside on the client, but could be shifted to a considerable extent to the server. Indeed all that the client really needed to provide was a user interface, and perhaps some local processing, which could be delivered perfectly well through a web browser. Andreessen therefore started talking about an 'Internet Operating System' based around Netscape Navigator which could run on Windows 3.1, Windows 95, an Apple Macintosh or even Unix. As Andreessen explained:

> The only difference technically between Netscape's Navigator browser and a traditional operating system is that Navigator will not include device drivers. (Moeller & Dodge, 1996)

Such talk, which turned Windows into little more than a "poorly debugged set of device drivers" (as Andreessen tactfully put it) was music to the ears of a number of companies unhappy with Microsoft's growing dominance. Sun Microsystems, whose products offered an alternative to both Windows and the PC architecture, had developed a computer language called Java which created programs designed to run within a software construct called the Java Virtual Machine (JVM). A JVM could be created for any operating system, which led to Java being promoted as the language that let you "write-once-run-everywhere" (Sun Microsystems, 1996). Java was unveiled at SunWorld 95 in San Francisco, where Netscape took the opportunity to announce that it would be licensing the technology so that its browser could run programs written in Java, regardless of the underlying operating system.

Oracle supported the concept, as it was not keen on the moves that Microsoft was making into its market. Likewise, Windows NT threatened Novell's network operating system NetWare. As Dave Kearns put it:

> Each of these products, on its own, is better than the comparable Microsoft offering – absolutely better or better because of its ability to run across different platforms. But it remains to be seen if this revitalised lineup can stop or even slow Microsoft's seemingly inexorable drive to dominate all facets of computing. (Kearns, 1997)

Given IBM's previous battles with PC clone manufacturers, it was perhaps inevitable that this group of companies became known as the 'Gang of Four'. However there the similarities end. Although there were attempts to build a set of products around the concept of an Internet Operating System, such as Oracle's Network Computer initiative which was endorsed by a number of manufacturers including both IBM and Apple, they came to little, as the hardware was too expensive and for most people the Internet was simply not fast enough. However this was a conflict that would be resolved in the law courts rather than in the marketplace.

Microsoft had already been accused of anti-competitive practices by a number of its competitors, which resulted in the US Department of Justice opening proceedings against the company in August 1993. This led to a judgment the following July in which Microsoft agreed to a number of restrictions on the way in which it licensed its software to PC manufacturers. In particular, Microsoft agreed not to make the licensing of one product, such as Windows, conditional on the licensing of another, such as its Office suite. Unfortunately the phrasing of the agreement was ambiguous, as it also stated "that this provision in and of itself shall not be construed to prohibit Microsoft from developing integrated products" (USA v. Microsoft, 1994), a clause that was to cause no end of problems.

13.   Through the looking glass

For most, the establishment of Windows and the PC compatible as a de facto standard was a good thing. For users, it meant that most of the computers they encountered ran the same applications as their own and could read their disks and files. For software companies that were writing applications, it made it easier to reach their customers – at most they'd have to write two versions of each application: one for Windows and one for the Apple Macintosh. It was a considerable improvement over the 1970s, when they were faced with a host of incompatible platforms, or the uncertainty surrounding OS/2 that followed. However it was bad news for companies such as Digital Research, IBM, Netscape or Novell which had products that competed directly with Microsoft. As far as they were concerned, Microsoft was operating as a monopoly, and the US had laws to prevent monopolies from abusing their position.

In October 1997, following extensive lobbying from a number of companies, the US Department of Justice (DoJ) filed a petition holding Microsoft in contempt of its earlier judgment by requiring PC manufacturers to install Internet Explorer alongside Windows 95, and requested that Microsoft be fined an unprecedented one million dollars a day until it desisted. Predictably, Microsoft's response was that Internet Explorer was an 'integrated product' and so not prohibited by the judgment. As Bill Gates put it:

> "There's no magic line between an application and an operating system that some bureaucrat in Washington should draw. It's like saying that as of 1932, cars didn't have radios in them, so they should never have radios in them. (Heilemann, 2000)

Gates argued that the government was in effect requiring Microsoft to stop innovating, and by doing so was acting on behalf of Microsoft's competitors rather than its customers.

The case was heard by Judge Thomas Penfield Jackson at the US District Court for the District of Columbia (home of the country's capital, Washington DC). Jackson agreed to waive the fine for the time being, on the understanding that Microsoft would comply with a preliminary injunction by offering PC manufacturers a version of Windows that did not come with a web browser until a final judgment could be made. Realising that to do so would undermine its case, Microsoft's response was to offer a version of Windows 95 in which all of the files used by Internet Explorer had been removed, including those vital to the workings of the operating system itself – a response that could at best be described as childish. Jackson, who was becoming increasingly exasperated with Microsoft's tactics, responded rather mischievously by claiming that it had taken him "less than 90 seconds" to remove Internet Explorer from his computer without disabling Windows 95, adding, "If it's not that simple, I'd like to have it refuted" (Angwin, 1998).

Microsoft's appeal against Jackson's decision was heard in April 1998. It was a tough one for Microsoft, as the DoJ was able to present considerable evidence of Internet Explorer being a product "separate and distinct" from Windows 95. For a start, it was perfectly possible for a Windows 95 user to install and use a browser other than Internet Explorer. Microsoft itself distributed versions of Internet Explorer as a distinct product, both for Windows 3 and for the Apple Macintosh. There was even the evidence of Microsoft's own witness, who explained that Microsoft had actually made it possible to uninstall Internet Explorer 3 from Windows 95 in response to feedback from corporate customers who did not allow their employees to "access the Internet and spend all their time surfing the Web" (Microsoft v. USA, 1998).

In the event, Microsoft won on a technicality, in that the Court of Appeals agreed that Jackson had not given Microsoft enough time to comply with the preliminary injunction, which in any case the DoJ had not requested. Unhappy with the outcome, the DoJ responded by suing Microsoft for violations of the Sherman Antitrust Act, which put the whole matter on a more serious footing. This was the Act that had resulted in Thomas Watson, who became the first president of IBM, receiving a prison sentence in 1912.

All of this was highly demoralising for those who worked at Microsoft. One of the attendees at an annual retreat held for Microsoft executives in 1999 sums up the concerns of many at the time:

> ... we're in a crisis here and we need to address what we stand for as a company ... We've been called evil; most of us with outside friends and family are being questioned by them, asked why we're working for Microsoft if it's an evil company. (Arthur, 2012, p. 18)

The trial was also taking its toll on Gates himself. As his father told *Newsweek*, "It was in the press every single day. His own government, suing him, that's not chocolate sundae! He was concerned, he was angry, he was distracted from things he'd rather be doing" (Levy, 1999).

However Microsoft's business strategy was clearly working, in that Internet Explorer was steadily gaining market share. In August 1995, when Microsoft launched Windows 95, Netscape Navigator accounted for 80 per cent of the browsers in use. Six months later, Internet Explorer could claim 22 per cent, increasing to 32 per cent by the end of its first year (Penenberg, 2009, p. 54). By the start of 1998, Netscape was reporting that it had lost $88 million over the previous three months (Liu, 1998).

That said, it should be noted that Netscape's losses were not solely down to Microsoft business practices. In 1996, Netscape embarked on an ambitious plan known as Netscape ONE (Open Network Environment) in a bid to provide an alternative to the solutions that Microsoft was offering, both on the client and on the server. As a consequence, Netscape Navigator was re-packaged in 1997 as a component part of Netscape Communicator 4, where it was joined by an email client, a news client, an address book, a conferencing system and facilities for a shared calendar.

One of the new features offered by this release of Navigator was limited support for Dynamic HTML (DHTML), a technology that allowed a web page to contain more than just text and images. With DHTML, a web page could also include program code that could be executed within the web browser, so allowing a degree of interaction between the user and the page. For example, a web page could check that the user has entered a valid credit card number before sending it back to the server for processing, and pop up an error message if not.

Unfortunately for Netscape, Microsoft's Internet Explorer 4 was released around the same time, and came with rather more comprehensive support for DHTML. It also proved somewhat closer to the definition that the World Wide Web Consortium (W3C) was working on. This, coupled with the confusion surrounding Netscape ONE – or "NONE", as it became known (Blundon, 1996) – did little to help Netscape's market share.

The trial commenced in October 1998, with Judge Jackson again presiding, and did not go well for Microsoft from the start. One of the DoJ's key witnesses was

Avadis Tevanian, who at the time was Senior Vice President of Software Engineering at Apple. In a break from convention, Judge Jackson questioned Tevanian directly:

> "From a technological perspective, ... what benefit, if any, is there, do you believe in integrating a browser as distinguished from bundling it with an operating system?" Less than none, replied Tevanian. "What you're telling me is, you don't think there is any benefit and there may be a detriment to the ultimate consumer?! That's right, replied Tevanian. "My final question: Is it possible to extricate your browser from the operating system without otherwise impairing the operation of the system?" Certainly, replied Tevanian. (Heilemann, 2000, p. 295)

A key witness for Microsoft was Richard Schmalensee, dean of MIT's Sloan School of Management, who was there to testify that Microsoft did not hold a monopoly with respect to Windows. However he was somewhat embarrassed to be confronted with an article he had written for the *Harvard Law Review* which suggested that "persistent excess profits provide a good indication" that a monopoly exists (Schmalensee, 1982). Microsoft had reported a profit of $4.49 billion for its previous financial year (Microsoft Corp., 1998). Schmalensee responded, "My immediate reaction is 'What could I have been thinking?'" A little later he attempted to defend his position by stating that the article "does not, the way it sits, present a good indication about my present views" (Chandrasekaran & Leibovich, 1999). But the damage was done.

Then there was the memo that Microsoft senior vice president Jim Allchin had sent his colleague Paul Maritz in Jan 1997 stating that Microsoft should start "leveraging Windows from a marketing perspective" in order to gain market share from Netscape:

> I am convinced we have to use Windows – this is the one thing they don't have ... We have to be competitive with features, but we need something more: Windows integration. If you agree that Windows is a huge asset, then it follows quickly that we are not investing sufficiently in finding ways to tie [Internet Explorer] and Windows together ... [Windows 98] must be a simple upgrade, but most importantly it must be a killer on OEM shipments so that Netscape never gets a chance on these systems. (Department of Justice, 1998)

Allchin was suggesting a deliberate strategy to install Internet Explorer as an integral part of Windows 98, not for technical reasons but in order to destroy Netscape's market share. For the trial, Microsoft had prepared a video that included a demonstration of 19 benefits that Microsoft claimed were made possible by the "deep integration of Internet technologies" demonstrated by the recently-released Windows 98. Stepping through each one, lead counsel for the prosecution David Boies questioned Allchin as to whether the same benefit couldn't be achieved by installing a retail copy of Internet Explorer 4 on a PC running Windows 95. Each time, but with increasing embarrassment, Allchin was forced to admit that Boies was right (Heilemann, 2000, p. 304).

There was also the testimony of Bill Gates himself, which had been filmed before the trial commenced. Gates is by nature an engaging public speaker, often exhibiting a strong sense of humour – perhaps not as charismatic as Steve Jobs, but certainly capable of commanding an audience. However in the film he appears inept, obtuse and at times deliberately evasive. At one point Boies questions Gates about an email that he had sent in 1997 which had been designated 'Importance: High'. The resulting exchange is worthy of a Monty Python sketch: (Department of Justice, 1998)

> Boies: "And you typed in here 'Importance: High'."
> Gates: "No."
> Boies: "No?"
> Gates: "No, I didn't type that."
> Boies: "Who typed in 'High'?"
> Gates: "A computer."
> Boies: "A computer. Why did the computer type in 'High'?"
> Gates: "It's an attribute of the email."
> Boies: "And who set the attribute of the email?"
> Gates: "Usually the sender sets that attribute."
> Boies: "Who is the sender here, Mr Gates?"
> Gates: "In this case it appears I'm the sender."
> Boies: "Yes. And so you're the one who set the high designation of important, right, sir?"
> Gates: "It appears I did that. I don't remember doing that specifically."
> Boies: "Right. Now, did you send this message on or about August 15, 1997?"
> Gates: "I don't remember doing so."

Judge Jackson is reported to have repeatedly shaken his head and laughed at such antics as the tape played. Microsoft lawyers later claimed that they had been led to believe that the recording would not be used in court, and so had not briefed Gates accordingly, but in the event they decided to limit the damage by not putting Gates himself forward as a witness.

As Microsoft's defence fell apart, repeated attempts were made to persuade the company to negotiate a settlement, but to no avail. Eventually, in June 2000, Judge Jackson felt that he had no choice but to order that Microsoft be split into two separate companies, one responsible for operating systems and the other for applications. His judgment made some provisions for how this might be achieved, specifically stating that the company responsible for operating systems should not have "any right to develop, license, or distribute modified or derivative versions of the Internet browser." The two companies would not be allowed to enter into any joint ventures, and any technical information transferred between them would have to be made public at the same time. Microsoft was even instructed to create a secure facility where other companies could examine not only technical documentation relating to Windows but even source code, with a view to ensuring that their applications functioned effectively. Microsoft was also instructed to let PC manufacturers modify the Windows boot-up sequence if, for example, they wished to offer users an alternative web browser. A Compliance Committee was to be established with wide-ranging powers to inspect internal communications until the judgment expired ten years later.

Looking back at the case a few months later, Judge Jackson commented:

> This is like the battle that ended the War of the Roses. That's the way I look
> at this case, like the fall of the House of Tudor. Something medieval. (Brinkley
> & Lohr, 2000)

Needless to say, Microsoft was already preparing its appeal, even as the industry debated the ruling. Such a judgment was not unprecedented, bearing comparison with the breakup of the communications giant AT&T in 1982. However many felt uneasy, sympathising with the sentiments of Charles Condon, who had withdrawn his support of the suit a few months after the start of the trial with the following statement:

> As Attorney General for the State of South Carolina, I joined other states
> over a year ago in the antitrust action against Microsoft ... My expressed purpose
> was to ensure that the citizens and the businesses of South Carolina would be
> protected from any harm that might come from one company gaining a
> monopoly over access to the Internet.
>
> Recent events have proven that the Internet is a segment of our economy
> where innovation is thriving. The merger of America Online with Netscape and
> the alliance by those two companies with Sun Microsystems proves that the

forces of competition are working. Further government intervention or regulation is unnecessary and, in my judgment, unwise.

Therefore, South Carolina is withdrawing from the Microsoft antitrust action. I can no longer justify our continued involvement or the expenditure of state resources on a trial that has been made moot by the actions of the competitive marketplace ...

Over the last year, it has become clear that the government's case has been about Internet competitors, not about consumers. The government's witnesses are either Microsoft's competitors or paid government experts. Consumers have not taken a leading role in this action. That's because there are no monopolies on the Internet. Anyone who has been Christmas shopping lately knows that we are seeing rapid innovation and falling prices in the technology industry. The consumers of South Carolina are benefitting from freedom and competition in the marketplace ...

I am taking this step on behalf of South Carolina because we believe in the free enterprise system. Innovation should be left to entrepreneurs, not to government bureaucrats or to the courts. As long as I am Attorney General of South Carolina, I will do everything I can to keep it that way. (Condon, 1998)

From the outside, it seems inconceivable that Microsoft should put up such an incompetent and even at times farcical defence. As Joseph Nocera reported in *Fortune* magazine:

There are moments in this trial when it feels as though we've just gone through the looking glass with Alice - when the world seems so upside down it makes your head hurt. (Nocera, 1999)

However, in Microsoft's eyes it did not have a case to defend: it clearly did not have a monopoly in the software industry as a whole, and it could hardly be called to account if its competitors made no attempt to compete with Windows itself. As Bill Gates put it at a meeting with the DoJ held in May 1998, some months before the start of the trial:

Give me any seat at the table - Java, OS/2, Linux - and I'd end up where I am. I could blow Microsoft away! I'd have programmers in India clone our APIs. If you were smart enough, you could do it. (Heilemann, 2000)

The court's Findings of Facts, published July 1999, did deal with this issue, but contended: "Translating this theory into practice is virtually impossible." It took this stance on the basis that doing so would involve developing an operating system that presented the same API (Application Programming Interface) as Windows, which amounted to thousands of procedure calls that changed with each version.

However the court could have insisted that Microsoft publish a full specification of its API; after all, software companies were already given the details that they needed to write applications that would run under Windows.

But Gates' main defence was that Microsoft had the right to decide for itself what features and functions its operating systems offered. This sounds reasonable, but such a stance does raise the question of where Windows as a platform ends and where the world of applications begins. It also begs the question as to what responsibilities, if any, Microsoft takes on as supplier of such a platform. Indeed I can recall a conversation with a software supplier in which Microsoft was likened to a bloated and bountiful sow, suckling a large number of voracious piglets. Each piglet enjoys a good living, but there is always the risk that the sow will roll over and squash it. If the piglet is lucky, Microsoft will buy it out; if not, then its business disappears overnight.

With reference to Gates' video testimony, it is also worth noting that many industry leaders appear ineffectual and even childish when challenged in a public forum. Witness Rupert Murdoch in front of the UK government's 2012 inquiry into phone hacking, or executives from Starbucks, Amazon and Google being grilled by British MPs about their tax affairs.

In the event, the appeal court overruled much of Judge Jackson's judgment the following year, primarily on the basis of his conduct during and immediately after the trial. To quote the ruling:

> The trial judge engaged in impermissible ex parte contacts by holding secret interviews with members of the media and made numerous offensive comments about Microsoft officials in public statements outside of the courtroom, giving rise to an appearance of partiality. (Brick, 2001)

According to their Code of Conduct, judges should not comment publicly "on the merits of a pending or impending action", and that includes cases that are under appeal. Despite this, it became apparent that Jackson had held several off-the-record meetings with journalists, even before publishing his judgment. At one meeting, he told reporters that he found the testimony of Bill Gates "inherently without credibility." In another, with reference to the numerous embarrassing emails that came to light, he likened Microsoft to a drug cartel in that "they never figure out that they shouldn't be saying certain things on the phone" (Schwartz, 2001). However, although recognising that he had clearly breached the Code of

Conduct, the court came to the conclusion that it was only in his decision to break up the company that Jackson had actually demonstrated bias. The appeal court therefore disqualified Jackson from the point at which he ordered the breakup but, despite protestations from Microsoft, upheld most of the Findings of Fact and Conclusions of Law that had led him to do so. Although much discussion inevitably followed, it was at least a verdict from which most of the parties involved could take some comfort.

The appeal court therefore rejected Microsoft's contention that the monopoly held by the Windows operating system was the result of natural market forces, pointing out that many of Microsoft's actions only made sense when viewed as unlawful attempts to protect its monopoly. However it did recognise the inevitable irrelevance of the judicial process to such a fast-moving industry:

> What is somewhat problematic, however, is that just over six years have passed since Microsoft engaged in the first conduct plaintiffs allege to be anticompetitive. As the record in this case indicates, six years seems like an eternity in the computer industry. By the time a court can assess liability, firms, products, and the marketplace are likely to have changed dramatically. (USA v. Microsoft, 2001)

The appeal court returned the case to the District Court with the stipulation that a new judge be appointed to decide on a suitable remedy, and at this point the DoJ announced that it was no longer pursuing the breakup of the company. In response Microsoft, realising that it too would have to make concessions, drafted a proposal for settlement. The country and the economy were reeling from the terrorist attacks of 11 September 2001 which had taken place only a few months earlier, and conciliation seemed more appropriate.

Judge Colleen Kollar-Kotelly published her final judgment in November 2002, and this time the emphasis was on prohibiting Microsoft from retaliating against PC manufacturers which chose to offer a choice of operating systems to their customers, or include software that competed with features offered by Microsoft (such as an alternative web browser). She also mandated that Microsoft publish the Windows API, and include tools allowing manufacturers and users to replace Microsoft programs and utilities, such as the web browser or media player, with alternatives.

In order to enforce the ruling, Judge Kollar-Kotelly ordered that a three-person independent technical committee be formed with a dedicated office inside

Microsoft's headquarters in Redmond, Washington and access to anything, or indeed anybody, that it needed to ensure compliance. In addition, Microsoft should appoint its own Compliance Officer with responsibility for ensuring that employees were aware of and conform to the ruling. The ruling was to stand for five years, with the provision for a two-year extension in the event that Microsoft was found to be "engaged in a pattern of wilful and systematic violations."

However this was far from the end of the matter, as attention turned to Judge Jackson's Findings of Fact, a document of over 200 pages which, having established that Microsoft did indeed enjoy a monopoly, went on to detail the steps which Microsoft had taken to hinder the activities of its many competitors, who were now looking for compensation. There were also nine states that did not accept the settlement and were continuing with class actions on the basis that their citizens had been overcharged for Microsoft products.

In an attempt to put the matter to bed, Microsoft offered to provide some 12,500 underprivileged schools with software, services, training and cash worth a total of $1 billion, together with a million copies of Windows to install on refurbished PCs. As a spokesman told the media:

> Microsoft really went the extra mile in trying to put together this settlement proposal. We thought it was creative, we thought it was a way to resolve litigation and bring some real benefits to the neediest school children in the country.

However it was quickly pointed out – most notably by Apple – that such an arrangement would also serve to increase Microsoft's hold on the market. Instead Apple suggested that Microsoft pay the $1 billion in cash to an independent foundation whose members could spend it on the computer equipment of their choice. The debate was resolved by US District Judge Frederick Motz, who rejected Microsoft' offer, largely because he felt the proposed amount was not enough.

Between May 2003 and October 2005, Microsoft was to pay out well over $5 billion in settlement of the various cases arising from the DOJ's Findings of Fact, including $750 million to AOL (which now owned Netscape), $605 million to the European Commission for breaching European competition laws, $1.95 billion to Sun, $536 million to Novell, $150 million to computer manufacturer Gateway, $850 million to IBM, and $761 million to RealNetworks, whose products competed with the media player technologies that Microsoft had also integrated into the Windows operating system (Rosoff, 2005).

These are huge sums, but to put them in perspective, Microsoft's annual revenue rose from some $28 billion in 2002 to over $44 billion by 2006, while its net profit more than doubled over the same period, from $5.4 billion to $12.6 billion (Microsoft, 2006). Furthermore, to quote a briefing published by the American Antitrust Institute in August 2003:

> Microsoft now has a 95 per cent of the share of OS market and 96 per cent of the Office Suite market. Most tellingly, given that the U.S. filed suit to prevent Microsoft from monopolizing browsers, today Microsoft has a 94 per cent share of the Internet browser market. (Hawker, 2003)

Microsoft had settled accounts with its competitors, and the threat posed by the 'Internet operating system' seemed to be over. Almost every PC compatible across the world was running Windows, now available in powerful 32-bit versions, and almost every business was using Microsoft Office. Microsoft's position seemed unassailable.

Ironically, the most noticeable outcome of all this legal activity for the ordinary user is the proliferation of often useless third-party software that assails you the first time that you turn on a new PC, put there by hardware manufacturers in an effort to get you to sign up to their offerings. In contrast, Macintosh owners continue to benefit from Apple's undiluted ownership of both hardware and software, which results in a smooth and seamless experience.

14. Saving Apple

The company that John Sculley inherited following the departure of Steve Jobs in 1985 was a company in chaos; a company at war with itself. On one side was the division responsible for the Apple II, which was still selling like hot cakes but feeling unloved. On the other was the division responsible for the Macintosh, leading the world in technical innovation but falling woefully short of its sales targets. Over in the PC world, sales were growing steadily as economies of scale and fierce competition drove prices down, but this was leaving Apple with a smaller share of the market – falling from almost 20 per cent in 1983 to just 11 per cent by 1985 (Carlton, 1997, p. 20).

On the other hand, Microsoft had yet to release Windows, and indeed could still be viewed as an ally. The Macintosh itself represented the state of the art, demonstrably more advanced than DOS-based PCs with their primitive text-based command lines. Apple could not afford to sit on its laurels, but it had some breathing space.

Sculley's response to the tensions within the company was to reorganise, dissolving Jobs' product-based divisions into a horizontal structure with one division in overall charge of marketing, another in charge of research, another manufacturing, and so forth. The theme was one of unity under the banner of 'One Apple', but the company was deeply traumatised by the departure of Steve Jobs, with some employees donning t-shirts that proclaimed, 'We want our Jobs back' (Rose, 1989, p. 311). This was the situation that Jean-Louis Gassée faced when he was put in charge of product development, which meant managing both engineering teams, including many people who had worked closely with Steve Jobs himself.

Then there was the question as to the direction the company should take now that Jobs was gone. Gassée favoured a 'high-right' strategy, referring to the sector of a price/performance graph currently occupied by the Macintosh. He argued that the company should maintain a high profit margin so that it could fully fund the research and development necessary to deliver the performance that would justify the price. This was why he rejected the suggestion put forward by Bill Gates that

Apple should license Macintosh technology to other companies (see page 141). Other board members disagreed, arguing that Apple should attempt to increase its market share by reducing prices; a view that Sculley shared, although he also suggested that the Macintosh be aimed at the corporate market, which would be prepared to pay more for the right product. This strategy also required machines that occupied that 'high-right' sector, so for the time being Gassée prevailed. It would also require machines that could be connected to the PC compatibles that already dominated the corporate desktop.

The first fruit of the new regime was the Macintosh Plus, which Apple announced in January 1986 at the Appleworld Conference in San Francisco. The Mac Plus came with a rather more useful 1MB of memory, which could be expanded to a massive 4MB, and Gassée took the opportunity to break with Jobs' concept of the Macintosh as a self-contained 'appliance' by including a SCSI socket, allowing users to connect a range of high-speed peripherals such as hard disk drives and image scanners. It was received well. As a recent retrospective put it:

> Before the Plus, people spoke of the Macintosh in terms of its potential – as it could be. After the Plus, people spoke of the Mac as it was. (Edwards, Happy 25th Birthday, Mac Plus, 2011)

Prices started at $2,599, and Apple took the opportunity to reduce the 512K Macintosh to $1,999. At the same time, Apple announced a future generation of machines that would offer an even more 'open' architecture, with slots that could accept expansion cards made not only by Apple but also by other manufacturers. As one analyst commented, "This is closer to IBM's strategy with the PC, XT and AT" (Ranney, 1986).

The promised machines appeared just over a year later. The Macintosh SE, which became available in March 1987, was very similar to the Mac Plus except for the provision of a single expansion slot (the 'SE' stood for 'System Expansion'). The design of the slot was very basic, essentially giving an expansion card direct access to the microprocessor. The machine started at $2,899, or $3,898 with a 20MB hard disk, and was frequently used with an Ethernet card so that it could be connected to a PC network. It was also the first Macintosh to come with a cooling fan – something that Jobs had rejected because of the noise it made.

The SE was followed in May by the Macintosh II, a flagship product that firmly occupied that 'high-right' corner. This was based around Motorola's 68020 processor, a 32-bit successor to the 68000 used in the Plus and the SE (and in the original Macintosh). It came with 1MB of memory as standard but could be expanded to an unprecedented 20MB. The basic machine came with no display or graphics capability but instead boasted slots for six expansion cards that conformed to the rather more advanced NuBus specification which became standard on future Macintosh machines. Into this could be plugged Apple's first colour graphics card, capable of resolutions up to 640 by 480 pixels in either 16 or 256 colours.

A usable configuration with a monochrome display cost just under $5,000, while a colour display pushed the price to more than $5,500. High prices indeed, but this was way ahead of anything the PC world could offer. Software developer David Winer summed it up for *InfoWorld* at the time:

> All the standards that made the Macintosh Plus so wonderful are now bigger and better and more powerful. We're pushing the outside of the envelope again. This is the 'infinity machine' we've been waiting for ever since we filled all 640K of the IBM PC.
>
> From this point on, no one can say that the Macintosh is a toy. It has become the most powerful personal computer on the market ... (InfoWorld staff, 1987)

Even Alan Kay, the man who conceived the Dynabook back in 1972, displayed some enthusiasm:

> ... in my opinion, the Macintosh II is the first microcomputer that's actually good. Mind you, it's still a subset of the stuff we did at PARC – it takes us to about the level we reached at Xerox in 1978 to 1980, something like that. (Webster, 1987)

As the desktop publishing revolution got underway (see page 130), Gassée's strategy began to really deliver. By the end of 1987, annual sales revenue had grown to almost $2.7 billion. The profit margin on the Macintosh range was growing too, and by 1988 exceeded 50 per cent. Money was pouring in, and the company was only too happy to spend it. Sales representatives could win 'Golden Apple' awards which included luxury cruises in the Caribbean. Engineers received generous stock options and automatic 10 per cent pay increases every six months. Sculley himself became one of the highest paid chief executives in the country.

Money was pouring into research and development too, at a rate of over $200 million a year. However it was not being spent wisely. Gassée had worked in the industry for nearly 20 years, but he was reluctant to interfere with the laissez-faire environment in which the engineers worked which had proved so productive in the past. Paul Mercer, who worked under Gassée at the time, recalls:

> He was very cool, very classy, and he gave engineers room to do stuff, but he wasn't really adult management. (Carlton, 1997, p. 83)

Or as Jim Carlton put it in his history of the company through this period: "The patients were running the asylum" (Carlton, 1997, p. 84).

The situation wasn't helped by Sculley's propensity to reorganise on a regular basis. Initially Gassée benefited from these reorganisations, first gaining control of product marketing as well as development, and then being made president of a new Apple Products division which added manufacturing to his responsibilities. His star was clearly rising, but corporate politics and a lack of consensus were taking their toll. A statement made by Sculley, when challenged that he lacked a coherent strategy, is particularly telling:

> I had a very consistent strategy with Apple that never changed. Using my marketing experience, I wanted to build a great brand, through advertising and public relations ... Secondly, because Apple was a feudal society and had been led by this mesmerizing cult leader [Steve Jobs], I did not know how to lead Apple from inside the company. My way of doing it was to lead outside the company [through speeches and external meetings], following distinct themes: education and multimedia. (Carlton, 1997, p. 75)

An example was the Aquarius project. A concern for Apple at the time was its reliance on Motorola for the microprocessors used by the Macintosh range. Jobs' decision back in 1980 to use the 68000 for first the Lisa and then the Macintosh made sense at the time; and its successor the 68020, as used in the Macintosh II, was easily the match of the Intel 80286 used by IBM for the PC AT. However PC compatibles based on Intel's 80386 were beginning to appear and Motorola had yet to release its third generation processor. Despite warnings that it would put them in competition with Intel and Motorola, both of which had invested billions in development and production, Gassée became convinced that Apple should design its own microprocessor. The project was initiated in 1986, and Gassée bought the team a $15 million Cray supercomputer to help with the design work.

Over the next few years the project spiralled out of control. At one point there was even talk of putting four processor cores on a single chip – something that was only implemented by Intel some thirty years later. Sculley finally terminated Aquarius in 1989, by which time it had cost around $20 million. Al Alcorn, who was briefly in charge of the project, commented:

> Aquarius was an example of the willingness of the management of Apple to embark on grand projects that skilled technical management would have known was unfeasible. (Carlton, 1997, p. 88)

Then there was the 'chip famine' of 1988, a reduction in the supply of memory chips from Japanese manufacturers caused in part by the growing popularity of the personal computer, but also blamed on a trade agreement intended to bolster American semiconductor manufacturers against their Japanese competitors. As a result, the cost of memory chips more than doubled over just a few months (Olmos, 1988).

Stiff competition amongst PC manufacturers meant that most went some way towards absorbing the increase in cost. However, anxious to maintain profit margin, Gassée continued purchasing chips unabated, passing the additional cost on to Apple customers by raising the price of the Macintosh itself – in some cases by almost 30 per cent. Unimpressed, many chose to buy their Macintoshes with as little installed memory as possible, which they then topped up with chips bought elsewhere. Sculley reversed the price rise a few months later, but Apple was left with a substantial stock of expensive memory chips and had to announce that its earnings for the first quarter of 1989 would be 42 per cent lower than the year before. The company's arrogance had tarnished its image with both customers and shareholders.

There was also the Macintosh Portable. Even by 1987, the demand for a portable version of the Apple Macintosh was apparent, to the extent that a number of companies were already selling them. Colby Systems, for example, sold the Lap-Mac for $4,995, which included a gas-plasma display, 1MB of memory and a single floppy disk drive. Like the similar Dynamac from Dynamac Computer Products, it weighed over 7kg.

Apple was not prepared to sell its motherboards (the electronic circuit boards that made the Macintosh unique) as separate items, so to get hold of the electronics, these companies purchased complete Macintosh Plus computers and took them apart. Dynamac Computer Products had signed a 'value-added reseller' agreement with Apple that allowed it to purchase Macintosh Plus computers at a reduced price, and Apple made it clear that it did not regard the deal as exclusive. However Colby Systems took a different approach, trying to persuade Apple to supply it with just the motherboards, and in the meantime going elsewhere to buy complete machines. In the end, Colby was selling its Walk-Mac SE in unfinished form to authorised Apple dealers who would complete construction using motherboards extracted from their unsold Macintosh SEs: not a solution of which Steve Jobs would have approved.

Gassée responded to the appearance of these machines by saying, "You will not see a 'flat' Mac from Apple anytime soon, which is not admitting or denying our interest in the subject matter" (Speigelman, 1987). In the event, when the Macintosh Portable finally did appear two years later, it proved a disappointment. It weighed as much as a Dynamac or a Lap-Mac, but had a starting price of $5,799 for a basic configuration. It did boast a display larger and sharper than any other portable at the time, and a battery life of up to 10 hours (thanks to its heavy lead-acid battery), but it was competing with PC compatible portables that weighed less than 5kg and cost under $3,000. Apple engineer Paul Mercer remembers an employee coming on stage to collect the Mac Portable that she had won in a company raffle: "She almost dropped it because it was too heavy" (Carlton, 1997, p. 104). As a result of such incidents, it became known as the 'Luggable', coming 17th in a 2006 retrospective of 'The Worst Tech Products of All Time':

> Some buildings are portable, if you have access to a Freightliner. Stonehenge is a portable sun dial, if you have enough people on hand to get things rolling. And in 1989, Apple offered a "portable" Macintosh – a 4-inch-thick, 16-pound beast that severely strained the definition of 'laptop', and the aching backs of its porters ... Some computers are affordable, too; the Portable met that description only if you had $6,500 of extra cash on hand. (Tynan, 2006)

To be fair, the IBM PCjr, which had come out four years earlier, was four places ahead of it. Furthermore, the Mac Portable could boast of being the first machine to send an email from space, when it was used by the crew of the Atlantis space shuttle to send the message: "Hello Earth! Greetings from the STS-43 Crew. This is the first AppleLink from space. Having a GREAT time, wish you were here, ... send cryo and CS! Hasta la vista, baby, ...we'll be back!" (Apple, 2012)

By early 1989, following the drop in earnings and the resulting decline in share price, there was a growing feeling that Apple was losing direction. Furthermore, IBM had already launched OS/2 with a graphical user interface in the form of Presentation Manager, and Microsoft Windows was beginning to look like a viable proposition. There was some way to go yet, but the Macintosh's position was not looking quite so unassailable. As Richard A. Shaffer, editor of Technological

Computer Letter, put it, "I wish there were more science and less science fiction. I don't have any sense from Apple that there is the great vision from the top." Gassée in particular came in for criticism with complaints that he "... has gotten as far as he has because he rhapsodises philosophical with a French accent and wears a single diamond earring." Sculley supported Gassée publicly: "He loves to be very French; he loves to bathe his language in a thick French accent. He's controversial, but he's exciting." (Lazzareschi, 1989). However behind the scenes, the tension was growing.

Christmas saw further falls in profits and share price, and Sculley started to think about cutting costs. Finally, the two of them slugged out their differences over what Gassée described as "the Last Supper" at the fashionable Maddalena's in Palo Alto. Sculley maintained that Apple had to increase market share, but Gassée insisted that that the only way forward was to maintain the 'high-right' position that differentiated the company from the competition. Sculley responded by adding the role of CTO (Chief Technical Officer) to his own existing roles as CEO, Chairman and President of Apple, and appointed Michael Spindler as COO (Chief Operating Officer) in charge of manufacturing and product marketing, so depriving Gassée of many of his responsibilities. Despite protests from the engineering team, Gassée resigned, negotiating a severance payment of $1.7 million in the process (Carlton, 1997, pp. 128-130).

Apple had progressively improved the Macintosh operating system over the years. System 5 had been introduced in October 1987 and included MultiFinder, which allowed it to store more than one application in memory at a time, displaying each on screen in a separate window. Users could switch from one application to another with the click of the mouse, and copy and paste information between them. They could also choose not to run MultiFinder, allowing the hardware to devote all of its attention to the needs of a single application, which made the Macintosh particularly suited to time-critical applications such as those devoted to music or film. System 6 was released in April 1988 and brought support for the 68030 processor. By the time that Microsoft released Windows 3.0, Apple was working on more sophisticated projects that went by the codename 'Blue' and 'Pink' (so called because of the colours of the index cards used to jot down ideas during the original brainstorm).

Apple had a long-standing practice of demonstrating the superiority of the

Macintosh by placing it next to a PC compatible and running versions of the same application on each. Shortly after the launch of Windows 3.0, some six hundred Apple employees gathered in the Santa Clara Convention Center to assess its performance against a prototype of System 7. As Duane Schulz recalls:

> The engineers onstage proceeded to run programs side by side on two computers, and Windows had a lot of inconsistencies. Everybody would crack up. It was an attitude of 'We're better.' (Carlton, 1997, p. 133)

However Sculley was not so sure, recalling, "When I saw [Windows 3.0], I thought it was pretty damn impressive." The demonstration made him realise that the industry was changing:

> In 1990, for really the first time, Microsoft and Intel started to work together. Before, people in the industry thought the computer was the box. But Intel said, 'No, it's the processor.' And Microsoft said, 'It's the software'. Together, they could run the industry. Meanwhile, we, IBM, Compaq and the other hardware manufacturers thought we were the industry. (Carlton, 1997, p. 134)

Sculley realised that Apple could not fight such a combination by itself, particularly if it was going to cut costs. At the time, IBM was looking to reduce its dependence on Intel, and its relationship with Microsoft was falling apart. Sculley began to think the unthinkable – that Apple might strike a deal with Big Brother.

By modern standards, Intel's early microprocessors were fairly rudimentary. The 8088 and 8086 processors, on which the IBM PC and early PC compatibles were based, understood 118 separate instructions. Such instructions implemented basic operations, such as adding to numbers together, or fetching a value from a particular location in memory. However each successive design added to the instruction set. The 80286, for example, understood 142 instructions, while the 80386 understood over 200. Motorola was following a similar strategy with the 68000 range used by Apple for the Macintosh. The DBEQ instruction, for example, tested for a condition and, if the condition was false, decremented a counter and looped back to an earlier instruction. Implementing such an operation on an earlier microprocessor would have required a considerable number of separate instructions.

Increasing the number of instructions that a processor understands brings many benefits. It makes the programmer's job easier, for example. However there is a downside, in that implementing each new instruction takes up precious space on the chip surface, and it could be argued that the space is wasted if the instruction is rarely used. Furthermore, the implementation of each instruction is hard-wired into the silicon, and might not be optimal in every circumstance.

IBM had been working on processor design long before Intel or Apple even existed, and had experimented with what became known as Reduced Instruction Set Computer (RISC) processors in the 1970s. A RISC processor deliberately sets out to offer a smaller set of more primitive instructions that can be executed much more efficiently. Programming a RISC processor is more demanding, but the end result should be a faster program.

IBM's research led to the development of what it called its POWER architecture (as in Performance Optimisation With Enhanced RISC), and by the early 1990s the company was looking for partners, having realised that it could only make an impact if it extended the technology beyond its own product range. The result was a phone call from IBM president Jack Kuehler to Spindler at Apple and a meeting at a hotel near Dallas, also attended by Sculley, to discuss whether Apple would be interested in helping to build a new microprocessor around IBM's POWER technology.

The meeting went well. After all, this was a very different Apple to the company that had launched the Macintosh under Steve Jobs. Sculley came from a similar corporate background to Kuehler, and the two had much in common. As Kuehler put it, "Alliances don't work unless both sides need each other. We needed Apple, and Apple needed us." (Carlton, 1997, p. 149).

Apple was wary of becoming totally dependent on IBM for the supply of the new microprocessor. After all, IBM's own range of personal computers did compete with Apple's. They therefore agreed to approach Motorola as a third partner. Motorola saw this as an opportunity to develop a product that would help it to compete with Intel, and so the AIM (Apple, IBM and Motorola) alliance was formally announced in October 1991 with the intention of creating a new line of 32-bit microprocessors based on IBM's POWER technology and dubbed the PowerPC (the PC stood for Performance Computing). The first to appear was the PowerPC 601, which became available in early 1993 and proved highly competitive.

Rather less successful was Star Trek, the project that would 'boldly go where no Mac has gone before'. Another company unhappy with Microsoft's ascendancy was

Novell, which had purchased Digital Research in 1991 and tried in vain to establish DR DOS, now renamed Novell DOS, as an alternative to MS-DOS. In February 1992 Tom Rolander, who had been present at those early meetings between Gary Kildall and IBM which had led to IBM selecting Microsoft rather than Digital Research to supply the operating system for the PC (see page 76), approached Apple with the idea of creating a version of the Macintosh operating system that could run on a PC compatible – something that Apple and Novell could sell as an alternative to Windows.

Following a meeting between Sculley and Andy Grove, Chief Executive of Intel, work started on the top secret project that July, with the involvement of both Novell and Intel. By November, the Star Trek team were able to demonstrate a version of Apple's System 7 operating system running on the 80486 processor of a standard PC compatible and doing things well beyond the capabilities of Windows. The project was demonstrated to Apple's Executive Council in early December and received the go-ahead. This was despite the reservations of Fred Forsyth, head of manufacturing and hardware engineering, who argued that such a product would reduce demand for the Macintosh itself, as it would no longer be differentiated by its superior operating system, and could also upset the partnership that they had so recently established with IBM and Motorola. Then, right at the start of the new year, came the surprise announcement that Roger Heinen, one of the leaders of the project, was leaving Apple to join Microsoft. As budgets tightened, Star Trek was eventually abandoned in June 1993.

Apple did finally get portable computing right with the PowerBook. Introduced in October 1991, this range of portables proved highly successful, with sales worth $1 billion in the first year, helping to boost annual revenue to over $7 billion and increase Apple's market share to 8.5 per cent.

It was a great product, but behind the scenes was a growing realisation that it marked a watershed, and that the company would have to change if it was to survive the relentless growth of Windows and the PC compatible. Apple was effectively 'caught between a rock and a hard place'. As a hardware company, it was competing with the likes of Compaq and Dell, but unlike them had to fund the research and development necessary to differentiate the Apple brand. As a software company, projects like Star Trek and Pink were held back by the need to support the Macintosh, putting them at a distinct disadvantage against companies such as

Microsoft and Intel that were not tied to a particular brand of hardware.

The PowerPC would help to fend off the inevitable, but Sculley felt that securing the company's long-term survival required something rather more radical. He had long been attracted by Alan Kay's concept of the Dynabook as a truly personal and portable computing device (see page 3), and Apple did in fact have a team working on such a thing, dubbed the Newton in reference to the original Apple logo, which had depicted Isaac Newton sitting beneath an apple tree. Larry Tesler, who had worked with Alan Kay at Xerox PARC, was in charge of the project, but it was generally accepted that a viable consumer product was some way off.

Sculley appreciated this, but saw in the Newton a product that could reduce Apple's dependence on the Macintosh. He also saw it as something that would allow him to leave his mark on the industry. Steve Jobs would always be remembered for the Macintosh; perhaps the Newton could be his legacy. In January 1992, Sculley was asked to give the keynote speech at the Winter Consumer Electronics Show (CES) in Law Vegas. He was careful not to discuss the Newton project specifically, but he did describe a new kind of consumer product that he called a "personal digital assistant". The PDA, as it became known, would include communication facilities so that users could extract information from their computers while on the move, and send messages to each other. According to Sculley, it would transform the industry.

It was a powerful vision, but those working on the Newton were nowhere near ready to deliver or even discuss an actual product. Just as with the original Macintosh project, there was considerable tension between the features that the team wanted to include, such as a touch sensitive screen that could accept and decipher handwritten text, and the need to keep the price below the target of $1,000.

However behind the scenes, Sculley was feeling the strain. That April, the judge rejected a large part of the argument that Apple had put forward to back its four-year-old case against Microsoft (see page 154), already compromised by the agreement that Sculley had signed with Bill Gates back in 1985. A month later, in May 1992, Sculley informed the board of directors that he intended to leave the company before his tenth anniversary in April of the following year. In the discussions that followed, the board came to the conclusion that Apple could not survive alone, and that the only solution was to sell the company. Sculley agreed to stay on so that he could lead the search for a suitable buyer, and opened negotiations with a number of prominent companies, to no avail.

The next idea was that Apple itself be split into two: one company producing Macintosh computers in direct competition with the likes of Compaq and Dell, while the other concentrated on developing cutting-edge software. But then came the news that sales over the first three months of 1993 had been far lower than expected, particularly of the PowerBook which had been doing so well. Costs had to be reduced quickly, and plans were put into place to cut 2,500 jobs. By June, the board had decided that Sculley should also announce his retirement. In time-honoured fashion, he was offered the figurehead role of Chairman, but resigned in October 1993 following a two-month sabbatical and a multi-million dollar settlement (Carlton, 1997, pp. 199-219).

It was against this background that the Newton MessagePad was finally launched at the MacWorld event held that August in Boston. Anticipation was high, but there was a degree of scepticism. Bruce L. Claflin of IBM, for example, likened the Newton to Big Foot:

> Many people talk about it but sightings are rare. It was overhyped and created expectations from which not only that company, but every other company, has to step back. (Rebello & Arnst, 1992)

There were also doubts about the price. It was launched at $700 despite market research showing that a PDA would have to cost less than $500, and ideally nearer $300, if it was to appeal to the ordinary consumer.

In the event, reception was decidedly mixed. In particular, it was criticised for simply not being ready. As one reviewer put it:

> I'm left with the impression of a company that knew the product was not ready for shipment and, indeed, did not have a plan for introducing it in any organized fashion. Then it realized it had to choose between delaying the introduction for a second time ... and pretending it had its act together and introducing the thing anyway.
>
> ... I think Newton will be a killer product - when it is finished. But being innovative is no excuse for poor implementation. And the Newton introduction is one of the messiest, poorly planned product introductions I have seen ... (Alsop, 1993)

Its idiosyncratic handwriting recognition became the subject of a Doonesbury cartoon which depicted the device recognising "I am writing a test sentence" as "Ian is riding a taste sensation" (Karlgaard, 2013). It also featured in a Simpsons episode with school bully Dolph's handwritten note "Beat up Martin" becoming "Eat up Martha" (the MessagePad ends up being thrown at Martin instead).

Handwriting recognition did improve in later models, and the range remained on sale until February 1998, when Steve Jobs terminated further development. Nevertheless it remained popular well into the following decade, inspiring newsletters and user groups. Indeed the irony is that today it's Steve Job's iPhone that is remembered, rather than Sculley's Newton.

Michael Spindler, already President and Chief Operating Officer, was Sculley's obvious successor – after all, he had been managing the company's day-to-day operations for some three years. His first job was to oversee the job cuts, which did little to boost morale or endear him to the staff. T-shirts started appearing that sported the *Jurassic Park* logo modified to read "Jurapple Park: the ultimate reorganization is extinction" on the front and "John Sculley got $1.52 million and all I got was this lousy T-shirt" on the back. There were also talk of "Spindler's List", a particularly tasteless reference as Spindler had been born in Nazi Germany.

On the other hand, Apple's alliance with Motorola and IBM was at last bearing fruit. Apple launched the Power Macintosh range on 14 March 1994 (Albert Einstein's birthday) with a series of adverts boasting, "The future is better than you expected!" The range was not only impressive, but also relatively cheap – for Apple at least – with prices starting at $1,800 for a box boasting a PowerPC 601 processor running with a 66MHz clock, 8MB of memory and a 160MB hard disk (keyboard and monitor not included).

Switching processor mid-stream is a risky strategy, because software companies will be reluctant to invest in the time and energy needed to write for a new platform until it has proved itself, and customers won't buy into a new platform that can't run the applications that they need. Apple got around this chicken-and-egg problem by ensuring that the PowerPC version of its System 7 operating system included a very proficient Mac 68K emulator, so allowing users to run applications originally intended for 68000-based Macintoshes on the new machines.

Less convincing were attempts to persuade people that the Power Mac could run Windows software, thanks to an emulator supplied by Insignia Software. In fact SoftWindows could only run applications that had been written for older 80286-based PC compatibles, and not current Windows applications written for

80386-based machines. However whether this actually mattered is a moot point. As a commentator pointed out:

> The Mac users that I know are not interested in using Windows, and the Windows users are not interested in using the Mac. There has to be a compelling reason [to switch from Windows]. (Quinlan & Borzo, 1994)

The Power Mac did do well, selling some 145,000 machines within the first two weeks and around 600,000 by October. However these were primarily to existing Macintosh customers looking to upgrade. It did stop them considering a Windows machine, but it was too late to substantially grow Apple's share of the overall market:

> In the eyes of the world, the Mac no longer has an edge over the PC in terms of usability. Windows has closed the gap. (Baum, 1994)

In desperation, Apple resorted to another strategy that it had previously rejected. At its annual shareholders meeting in January 1994, Spindler told shareholders that Apple would be offering PC manufacturers the chance to sell clones of the forthcoming Power Mac. Discussions were opened with a number of the larger manufacturers but few expressed much interest. After all, they were already doing very well producing Windows machines and didn't want to risk confusing their customers with a separate line of Macintosh compatibles, or jeopardising their relationship with Microsoft. Eventually, at the end of the year, Apple was able to announce its first licensing contract, to a little-known Californian company called Power Computing.

When Spindler first announced its intention to licence the Power Macintosh, the company had boasted that the initiative would lead to the Macintosh accounting for up to 30 per cent of the market within just two or three years (Markoff, 1994). But this was not to be. By May 1995, Power Computing was selling clones of the Power Macintosh by mail order for less than the shop price of the originals. As chief executive Stephen Kahng explained, "Our strategy is more for less" (Los Angeles Times, 1995). Predictably, the result was not an increase in the Macintosh's share of the market, but a decrease in Apple's share as customers chose the cheaper clones over the Apple originals. Only a handful of other licences were granted, and the search for further licensees was abandoned.

One of Power Computing's strengths was its streamlined production system, highly efficient and cost effective from procuring parts right through to final delivery. Apple, by contrast, was burdened by a plethora of product lines which were difficult to fulfil and the cause of much confusion. Initially there was the Compact or Classic Macintosh, designated the original all-in-one box, and the high-performance Macintosh II. These were joined by the Macintosh LC, a label indicating 'Low-cost Colour' and applied anywhere it was felt appropriate. Then there were the high-end Quadra machines which replaced the Macintosh II and were in turn superseded by the Power Macintosh. These were balanced by the low-end Performa brand, although none of the machines in the Performa range were actually new. Instead they were existing machines bundled with software for home and small business use: the Performa 5200, for example, was a repackaged Power Macintosh 5200 LC. The Centris brand was briefly introduced to denote machines in the 'centre' of the Macintosh range, but quickly abandoned as this was considered too confusing. And then there were the portable PowerBooks.

Apple's situation did not improve through 1995, as cumbersome supply lines resulted in shortages and inefficiencies which contrasted badly with the streamlined production lines of Windows-based machines, honed by fierce competition between their manufacturers. Apple's share price started to drop and then, with the Christmas period approaching, the company announced that it might actually make a loss. The share price fell 15 per cent in just four days, and in January 1996, Apple did indeed admit to a loss of $68 million over the preceding quarter and further job cuts. Shareholders were calling for his resignation, but Spindler had a final card up his sleeve.

At some point in the latter half of 1995, talks had begun between Apple and Sun Microsystems about a possible buy-out. Sun had recently released its Java technology and was in discussion with Netscape and Oracle over the possibility of establishing the Internet Operating System which would render Windows redundant (see page 175). Apple had the know-how to build machines that could run it and a brand that, despite everything, was still held in high esteem. It looked to be a good match, provided Apple could accept the price that Sun was willing to pay for the company. Initial talks proved promising, but the alarming drop in share price spooked Sun into reducing its offer. The board convened, and it quickly became clear that the choice was between accepting Sun's offer and continuing alone with a new Chief Executive.

Gilbert Amelio, who had joined the board in 1994 while retaining his position as Chief Executive of National Semiconductor (one of Apple's principal suppliers), spoke out against the deal, and on 30 January 1996 was voted in to replace Spindler. Despite the company's losses, and the impending cuts, Amelio negotiated a deal that would pay him some $3 million over the next year, while Spindler received a settlement worth nearly $4 million (The New York Times, 1996). The company itself was in deep trouble, but those at the top saw no reason to stop helping themselves.

An immediate concern for Amelio was the Macintosh operating system. With Windows 95, Microsoft had closed the gap to the extent that any technical advantage the Macintosh might still have was not worth the price premium. Apple's System 7 had been around for some time now, but nothing from Apple's fragmented research and development projects was going to replace it anytime soon. The fabled Pink operating system project had been spun off as Taligent, a joint effort with IBM, but was going nowhere. Copland was another attempt at an operating system that had been announced in May 1994, with assurances of a release the following year, but by October 1995 Apple had to admit that it was unlikely to be ready before 1997. As Don Crabb put it:

> ... our friends in Cupertino have *done it again*! They have mismanaged a development project that only has the fate of the company and its loyal developers' lives hanging in the balance. If it weren't so serious, it would be hysterically funny. (Crabb, 1995)

In desperation, Amelio started looking for alternatives outside the company. One possibility was OPENSTEP, a version of the NeXTSTEP operating system developed by the company that Steve Jobs had set up after leaving Apple back in 1985. NeXTSTEP was in part based on the well-established Unix operating system.

Jobs had created NeXT to produce graphic workstations for use in higher education, a project which would allow him to pursue his dreams without the cost constraints of the mass market. The first to appear some three years later was the NeXT Computer. This was certainly state-of-the-art, coming as a stark matt-black cube containing the latest generation Motorola 68030 processor and 8MB of memory, and a price tag of $6,500. It also boasted a new operating system with a graphical user interface that used a version of Adobe's PostScript language to generate the screen. The display demanded a great deal of computing power, but uniquely guaranteed that what you saw on the screen really would match what you got on the $1,995 NeXT Laser Printer (Thompson & Baran, 1988).

At these prices, the NeXT Computer was not going to sell in large numbers but it did find its way into many universities. This is the computer that Tim Berners-Lee used at CERN to develop WorldWideWeb (see page 167). However by 1992 the company had started work on a version of NeXTSTEP that would run on the Intel 80486 processor, and in 1993 it stopped making hardware altogether and changed its name to NeXT Software.

NeXT was not Jobs' only interest. A few months before he left Apple, Alan Kay had introduced him to Ed Catmull, who managed the computer division of Lucasfilm with Alvy Ray Smith. The division was developing cutting-edge software and hardware for computer animation, and the two managers were looking for an investor to help them to purchase the division from George Lucas and establish it as a company in its own right. The deal was completed in January 1986 and the division was renamed Pixar Animation Studios. Jobs had invested $10 million of the $100 million that he had made when he sold his share of Apple, gaining in return a 70 per cent share of the new company.

The original intention was that Pixar would sell high-performance animation systems. However the division already had a reputation for creating top-quality animations, including sequences in *Star Trek II: The Wrath of Kahn* and *Young Sherlock Holmes*. In 1986, John Lasseter had put together *Luxo Jr.*, a two-minute film telling the tale of two table lamps that became the first computer-generated animation to be nominated for an Academy Award. As with so many of Jobs' ventures, the company was losing money but its products were highly acclaimed. Over the next few years, Jobs continued to inject cash in exchange for shares, helping to finance Lasseter's *Tin Toy*, which became the first computer-generated film to win an Academy Award. By 1991 he owned the company outright, having invested a total of $50 million, although he didn't take the CEO position until 1994.

The next step for Pixar was a full-length feature film, but this would require outside funding. Pixar's biggest customer was Walt Disney, and in May 1991 they struck a deal. The result was *Toy Story*, which premiered in November 1995. Although it didn't win an Academy Award, it was nominated in three categories, while Lasseter himself won the Special Achievement Award. *Toy Story* became the highest grossing film of the year, beating *Batman Forever*, *Apollo 13* and Disney's own *Pocahontas*, and taking over $361 million worldwide (Box Office Mojo). One week after its premier, Jobs took the company public, and his investment became worth $1.2 billion. Ten years later, in January 2006, Disney bought Pixar in a deal that turned Jobs into Disney's largest individual shareholder.

However Jobs always insisted that it was not about the money. As he told Robert X. Cringely in 1996:

> I was worth about a million dollars when I was 23 and over ten million dollars when I was 24, and over a hundred million dollars when I was 25, and it wasn't that important because I never did it for the money. (Jobs, 1996)

Or, to put it another way:

> Being the richest man in the cemetery doesn't matter to me. Going to bed at night saying we've done something wonderful ... that's what matters to me. (Zachary & Yamada, 1993)

Certainly, Jobs did live frugally by comparison with other CEOs such as Sculley or his friend Larry Ellison, co-founder of Oracle. On the other hand, he was clearly attracted to the creative opportunities that money made possible, and he was not happy until he had complete control of a project. He was also becoming very good at making it.

Jobs' relationship with Apple was similarly ambivalent. He was particularly scathing about Sculley:

> Sculley destroyed Apple by bringing in corrupt people and corrupt values. They cared about making money – for themselves mainly, and also for Apple – rather than making great products. Macintosh lost to Microsoft because Sculley insisted on milking all the profits he could get rather than improving the product and making it affordable. (Isaacson, 2011, p. 295)

Perhaps the relationship was best summed up by Ellison:

> It's as if Apple is an old fiancée from college that Steve met again at a 20-year class reunion. Steve is happily married now with children, and has a great life. When he meets his old girlfriend again, she's an alcoholic and is running around with a bad crowd and has made a mess of her life. Even so, in his mind's eye, he still sees the beautiful woman he once thought was the love of his life. So what's he supposed to do? Of course, he doesn't want to marry her anymore, but he can't just walk away, because he still cares about her. So he puts her in a detox program and tries to help her meet a better class of friends and hopes for the best. (Schlender & Woods, 1997)

Discussions between the two companies were initiated by NeXT product manager John Landwehr who, having read that Apple was looking to acquire a new operating system, persuaded his colleague Garret Rice to contact Apple's chief technology officer with the suggestion that OPENSTEP might fit the bill (Markoff, 1996). Just a few weeks later, in December 1996, Amelio and Jobs were finalising Apple's purchase of NeXT Software for $400 million.

Amelio wanted Jobs to return to Apple in a managerial position as part of the deal, but Jobs resisted, as he didn't want to jeopardise his work with Pixar. Instead he took the part-time role of 'advisor to the chairman'. However, when Amelio introduced him as such at the January MacWorld conference in San Francisco, Jobs was welcomed with a standing ovation that lasted over a minute, and it soon became obvious that he would be in the driving seat. A few weeks later, the *Financial Times* was reporting:

> Mr. Jobs has become the power behind the throne. He is said to be directing decisions on which parts of Apple's operations should be cut. Mr. Jobs has urged a number of former Apple colleagues to return to the company, hinting strongly that he plans to take charge, they said. According to one of Mr. Jobs' confidantes, he has decided that Mr. Amelio and his appointees are unlikely to succeed in reviving Apple, and he is intent upon replacing them to ensure the survival of 'his company.' (Kehoe, 1997)

In June, following pressure from both the board and shareholders in the face of further losses, Amelio tendered his resignation. A month later, Jobs accepted a position on the board and agreed to help them to find a new CEO. For the time being, he still positioned himself in an advisory role, introducing himself at the August MacWorld as chairman and CEO of Pixar, although his keynote speech made it clear that he was once more in charge.

Another complicated relationship was that between Apple and Microsoft. The two companies were clearly in competition, both in the marketplace and in the courts, and it was a battle that Apple was losing. Furthermore, Jobs remained firmly of the opinion that Microsoft did not deserve to succeed, as Windows was essentially just a copy of the Macintosh. On the other hand, Microsoft had a soft spot for Apple, as its first applications had been written for the Macintosh. Microsoft made a bigger profit on applications such as Word and Excel than it did on Windows itself, and the Macintosh versions of these products were big sellers. According to one report, Microsoft Excel accounted for 90 per cent of the spreadsheet software found on the Macintosh, while 75 per cent of word processing was done using Microsoft Word (Salomon Brothers Inc., 1994). Apple would suffer badly if Microsoft withdrew its support.

Jobs therefore called Bill Gates, essentially asking him for help. Despite the decisions of the court, Jobs still maintained that Microsoft was in breach of patents held by Apple. However, as he later described the conversation, he continued:

> If we keep up our lawsuits, a few years from now [Apple] could win a billion-dollar patent suit. You know it, and I know it. But Apple's not going to survive that long if we're at war. I know that. So let's figure out how to settle this right away. All I need is a commitment that Microsoft will keep developing for the Mac and an investment by Microsoft in Apple so it has a stake in our success. (Isaacson, 2011, p. 324)

While Gates was not about to admit any patent breaches, he did believe that Jobs might be capable of reversing Apple's fortunes, and an agreement would allow Microsoft to benefit. Furthermore, Gates saw this as an opportunity to extend the reach of Internet Explorer, which would help in his battle against Netscape.

Jobs introduced the Microsoft deal to the Macintosh faithful at the August 1997 MacWorld conference in Boston, saying towards the end of his keynote speech:

> Apple lives in an ecosystem. It needs help from other partners. Relationships that are destructive don't help anybody in this industry. I'd like to announce one of our first new partnerships today, a very meaningful one, and that is one with Microsoft.

To the sound of gasps and groans from the audience, Jobs outlined the deal. Microsoft had agreed to pay $150 million for a six per cent holding in Apple which it would not sell for at least three years. Gates had also agreed to match every major release of Microsoft Office for Windows over the next five years with a comparable version for the Macintosh. All outstanding legal disputes would be cast aside, and all patents shared and cross-licensed for a period of five years. On Apple's part, Jobs had agreed to make Internet Explorer the default web browser for the next release of the Macintosh operating system – although, in an attempt to placate the audience, he quickly added that "the user can, of course, change their default should they wish to."

The crowd was then presented with a giant projection of Gates' face, beamed from his office in Redmond, looming large on the screen behind the lone figure of Jobs in an unintended parody of the Big Brother advertisement that had launched the Macintosh all those years ago. Unaware of the impact that he was having, Gates emphasised, "Some of the most exciting work that I've done in my career has been the work that I've done with Steve on the Macintosh." Once Gates had finished, Jobs admonished the audience:

> We have to let go of this notion that for Apple to win Microsoft has to lose ... I think if we want Microsoft Office on the Mac, we better treat the company that puts it out with a little bit of gratitude. We like their software. (Thomas, 1997)

The next MacWorld, held in January 1998 in San Francisco, saw the announcement of both Internet Explorer 4.0 for Macintosh and Mac OS 8.1, which included it as the default web browser but offered Netscape Navigator as an alternative. Internet Explorer remained the default choice until it was replaced by Apple's own Safari browser five years later.

Jobs had outlined a strategy for Apple several years earlier in a quote to *Fortune* magazine that had been published before Apple had even considered buying NeXT Software:

> If I were running Apple, I would milk the Macintosh for all it's worth – and get busy on the next great thing. The PC wars are over. Done. Microsoft won a long time ago. (Martin, 1996)

Finding himself in a position to do so, that is exactly what he did. He started by cancelling products like the Newton that he saw as a distraction, and terminating the agreements that allowed the manufacture of Macintosh compatibles – a decision that resulted in a payment of $100 million to Power Computing.

He then set about restoring the Apple brand, reiterating the original 1984 message in 1997 through a poster and TV advertising campaign dubbed 'Think Different' and featuring the faces of people like Albert Einstein, Mahatma Gandhi, Alfred Hitchcock, John Lennon and Pablo Picasso. The adverts were accompanied by the following text which was partly written by Jobs himself and read for TV by the actor Richard Dreyfus. The only reference to the company was the discrete display of the Apple logo:

> Here's to the crazy ones. The misfits. The rebels. The troublemakers. The round pegs in the square holes. The ones who see things differently. They're not fond of rules. And they have no respect for the status quo. You can quote them, disagree with them, glorify or vilify them. About the only thing you can't do is ignore them. Because they change things. They push the human race forward.

> While some may see them as the crazy ones, we see genius. Because the people
> who are crazy enough to think they can change the world, are the ones who do.
> (Rose G. , 2011)

Jobs also took the opportunity to rebrand himself as 'interim CEO', which was inevitably abbreviated to 'iCEO'. The title was not ratified on paper and he continued to draw no salary, officially remaining in an advisory role. However the board quietly dropped its search for a replacement.

Turning his attention to the company's production problems, he poached Tim Cook from Compaq, and in March 1998, Cook joined Apple as senior vice president for worldwide operations. Cook knew what it took to maintain efficient supply lines from his work at Compaq, and previously at IBM, and he set about applying his expertise to Apple:

> The company lost $1 billion in 1997 mainly as a result of asset problems,
> such as being too long on inventory. We had five weeks of inventory in the
> plants, and we were turning inventory 10 times a year. (Bartholomew, 2004)

A year later, Cook had brought the average length of time that inventory was stored down to just six days, comparable with the turnovers being achieved by PC clone manufacturers such as Dell. The process was painful, involving considerable loss of jobs, but it was necessary if the company was to survive.

Another part of the process was the rationalisation of Macintosh product lines. As Jobs stated in his keynote speech at the July 1998 Macworld event in New York:

> We went back to business school 101 and said 'What do people want?' We
> said, 'Well, they want consumer products and pro products.' And in each
> category we need desktop and portable products. If we have four great products
> that's all we need. (Kane, 1998)

For the professional there was the Power Mac and the Powerbook, and work started on new models powered by the third-generation PowerPC G3. For the consumer there would be two new ranges, also based on the G3 processor.

The English designer Jonathan Ive had joined Apple in 1992, moving from London to the company's offices in Cupertino. By 1996, Ive was running the design studio but feeling neglected. As he told Jobs' biographer:

> All they wanted from us designers was a model of what something was
> supposed to look like on the outside, and then engineers would make it as cheap
> as possible. I was about to quit. (Isaacson, 2011, p. 341)

This all changed with the return of Jobs to the company. Jobs had always appreciated the importance of design, and the two became close friends. The result was the iMac, which was launched in May 1998.

In many ways, the iMac was a return to Apple's Macintosh roots, coming as a self-contained unit that included the monitor as part of the box and even a handle at the top. Inside it was simply a re-packaged Power Mac, based around the same G3 processor. However it was also affordable, at $1,299 for a model that came with a very usable 32MB of memory, a 4GB hard disk, stereo speakers and a 15-inch colour monitor. Furthermore it had a colourful wrap-around case that was transparent so that you could see the electronics inside. Initially this was only available in 'Bondi Blue' but by the following January you could choose your colour from a selection of blueberry, grape, lime, strawberry and tangerine, with all models available in the UK for £779 plus tax. As Jobs said at the launch ceremony, "It looks like it's from another planet. A good planet. A planet with better designers." Or as *MacWorld* quipped, "It killed beige" (Edwards, 2008).

The iMac also introduced the 'i' prefix to indicate a device designed from the ground up to work seamlessly with the Internet – although it could equally be said to stand for 'individual' or 'innovation'. The iMac itself came with networking and a dial-up modem built in which, according to one advert, meant that you could get connected in just two steps.

The new strategy, and in particular the new iMac, worked a treat. Apple was able to claim 800,000 units sold within its first five months, which made it the top-selling computer in the United States. As Jobs put it at the January 1999 MacWorld Expo:

> During iMac's first 139 days, an iMac was sold every 15 seconds of every minute of every hour of every day of every week. (Apple Computer, 1999)

More significantly, a third of the sales were to people who had never bought a computer before, and 13 per cent were to people who had previously owned a PC compatible. By June 1998, Jobs was also able to claim sales of 750,000 of the G3-based Power Macs (Kane, 1998). Apple had turned the corner. Although the workforce had been halved, the company had turned a loss of over a billion dollars for the year to September 1997 into profits of $309 million for 1998, and over $600 million for 1999.

In January of the new millennium, Jobs finally accepted the official title of CEO, although remaining CEO of Pixar as well. He was also working on the next step of the strategy that he had outlined back in 1996. A year later, at Macworld in San Francisco, Apple released iTunes 1.0, a software program that allowed you to manage music stored in a digital format. The program was available for the Macintosh free of charge and was distributed with Mac OS X when it came out in March 2001. That October, iTunes was followed by the iPod, a tiny music-playing device that worked with iTunes so that you could store and play your music away from your computer.

Then came the online iTunes Store where you could buy individual songs for just 99 cents a time. The music industry was being hit hard by piracy, but after lengthy negotiations, first Warner Music and then other major music distributors signed up to the new service. Apple's iTunes Store sold a million songs within its first six days, and is now the biggest music vendor in the world. Apple sales for the year ending September 2012 were worth $156 billion, most of which came from the iPhone and the iPad. Macintosh computers accounted for less than 15 per cent. Jobs had achieved his vision.

As for the Apple Mac, at a conference held in San Francisco in June 2005, Jobs announced that all future models would be based around Intel processors, rather than the PowerPC, and took the opportunity to demonstrate a version of the Mac OS X operating system running on an Intel-based machine. By the following January, new MacBook Pro and iMac machines had appeared that were based around the Intel Core Duo processor, which was also to be found in top-range PC compatibles of the time. Indeed these machines differed little from current PC compatibles, which raised the question as to whether they could actually run Windows.

That thought inspired Colin Nederkoorn, who worked as a shipping broker in Houston, to purchase an Intel-based MacBook Pro to replace his desktop PC on the basis that it would be able to run both Mac OS X and Windows XP. What he really wanted was a machine that would offer him a choice of operating system when it was turned on, so he could choose between the two depending on what he needed to do. In order to make this happen, he initiated a competition, offering $100 to the first person to demonstrate an Apple Mac capable of doing so, and inviting others to contribute to the pot. The prize, which by then had grown to $13,854, was won in March 2006 by Jesus Lopez and Eric Wasserman, two programmers from the San Francisco region who delivered the necessary software patch for Windows XP together with a set of instructions.

Neither Microsoft nor Apple had any objections to the project, and indeed just two weeks later, Apple released its own solution in the form of a free utility called Boot Camp which went on to become an integral part of Mac OS X. As senior vice president Philip Schiller said: "We think Boot Camp makes the Mac even more appealing to Windows users considering making the switch."

Apple is somewhat less enthusiastic when it comes to attempts to run the Intel version of Mac OS X on anything but a Macintosh, and indeed the Mac OS X software licence explicitly states that it is not to be used on any "non-Apple-branded computer". However it can be done, with the result usually being referred to as a 'Hackintosh'.

Apple had dabbled with Intel processors before, with the ill-fated Star Trek project actually demonstrating a version of Apple's System 7 operating system running on an Intel 80486 processor back in 1992, well over a decade earlier (see page 196). However it took Steve Jobs to finally take the plunge. As John Sculley commented in 2010, just a year before Jobs' death as a result of cancer:

> I made two really dumb mistakes that I really regret, because I think they would have made a difference to Apple. One was when we were at the end of the life of the Motorola processor ... we took two of our best technologists and put them on a team to go look and recommend what we ought to do. They came back and said it doesn't make any difference which RISC architecture you pick, just pick the one you think you can get the best business deal with. But don't use CISC [as used by Intel].
>
> Intel lobbied heavily to get us to go with them ... [but] we went with IBM and Motorola with the PowerPC. And that was a terrible decision in hindsight. If we could have worked with Intel, we would have gotten onto a more commoditized component platform for Apple, which would have made a huge difference for Apple during the 1990s. Intel would spend $11 billion and evolve the Intel processor to do graphics ... So we totally missed the boat.
>
> The other, even bigger failure on my part was if I had thought about it better ... I would have said, "Why don't we go back to the guy who created the whole thing and understands it? Why don't we go back and hire Steve to come back and run the company?" It's so obvious, looking back now, that that would have been the right thing to do. We didn't do it, so I blame myself for that one. It would have saved Apple this near-death experience they had. I'm actually convinced that if Steve hadn't come back when he did – if they had waited another six months – Apple would have been history. It would have been gone, absolutely gone. (Sculley, 2010)

15.   Geeks bearing gifts

The final challenge to Microsoft's supremacy came not from another company or a new technology, but rather from a movement that was attempting to undermine the very nature of its business. Such a challenge was possible because, like music or literature, the value inherent in a computer program such as Microsoft Windows or Office lies not in its physical manifestation but in the 'intellectual property' that it gives us access to. We don't buy an audio CD in order to own a shiny piece of plastic – we buy it so we can listen to the music recorded on its surface. Once we've bought it we can copy the music onto a digital music player, for example, and happily discard the CD itself. The problem is that someone else can take your CD and do the same, so gaining access to the music free of charge.

Many see this as a 'victimless crime' because the owner of the CD has not been deprived of anything of value and still has access to the music. However the owners of the intellectual property – the musicians, writers or programmers – can argue that they have been deprived because the thief is enjoying the fruits of their efforts without paying for it. The thief may even be making money from their work by selling it on and pocketing the profit. This is why we have laws of copyright, limiting the 'right to copy' to the original author of a work.

In the early days of the music industry, copyright theft or 'piracy' was not much of a problem because it was simply not economical. People did copy vinyl records onto magnetic tape, but it was a laborious process and the records themselves were not expensive. The practice become more prevalent once audio CDs became available, primarily because the music industry took the opportunity to increase the price, which many felt was unreasonable. However it was the personal computer and the Internet that precipitated the crisis, replacing audio CDs with digital files that could be copied with ease and without degradation. The crisis only abated once services like iTunes Store, which gave easy access to music files for a reasonable price, became accepted by both the industry and its customers.

In the case of software, intellectual property comes in two forms. The most valuable is the source code of the program as written by the programmer in a programming language such as BASIC or C. For a commercial software company, the source code is the crown jewels. The source code of a program like Microsoft Windows or Office, for example, has a value that can be measured in billions of dollars.

However this is not what ordinary users need if they want to use these programs. Instead, what the customer is usually offered is a licence to install and run a copy of the compiled executable – the binary machine code that is generated from the source code – in accordance with certain conditions. The terms of the licence vary from company to company, but are likely to restrict installation of the program to just the customer's own machines, and to prohibit transfer of the licence to anyone else unless the software is first removed from the customer's computers. Such licences are referred to as 'proprietary' licences because the copy of the compiled program that is in the customer's possession remains the property of the company.

The software industry predates the microcomputer by several decades. IBM, for example, was making big money from programs like SABRE back in the 1960s (see page 5). However a lot of programming was done by researchers working within academic establishments such as MIT or Stanford Research Institute. Here a very different ethos prevailed, with programmers happy to let others access and improve upon their source code. It was generally accepted that the programmer's name should be attached to the sections of code that they had written, but computer programming was seen as a communal effort, benefiting from the contributions of others.

It was also a matter of scale. Software companies had been used to cultivating close relationships with a relatively small number of customers, but that all changed with the arrival of the microcomputer. As Jean-Louis Gassée recalls from when he joined Apple in 1981:

> I went to one of [Apple's] warehouses ... and I saw forklifts moving pallets of software. When you came from the mini-computer industry, the idea that you would sell pallets of software was a shock ... Apple was every month selling more copies of word processing software than Exxon [Office Systems] was selling of word processing systems in one year. (Gassée, 2012)

It was the clash of these two cultures that provoked controversy right from the birth of the microcomputer, when hobbyists started copying the BASIC interpreter that Bill Gates had written for the MITS Altair (see page 31). Gates defended his right to sell copies of his program in his 'Open Letter to Hobbyists', making the case that programmers had a right to be paid for their work, and that the industry needed professional programmers if it was to flourish. The most coherent response came from Jim Warren, who went on to edit *Dr. Dobb's Journal* which promoted software as something that should be distributed for free, charging only for the cost of distribution.

Somewhat ironically he also suggested that programmers wanting to make "significant sums of money" should sell their software to hardware manufacturers, which is essentially what Microsoft did, first with MS-DOS and then with Windows. By 2005, over 80 per cent of the revenue that Microsoft made from its Windows operating system came from PC manufacturers buying it to pre-install for their customers. Less than 20 per cent came from end-users buying it in a shop or as part of a corporate licence for their business (Microsoft Corporation, 2005).

The widespread adoption of proprietary licensing came about largely because the nature of the customer changed. Hobbyists and hackers love nothing more than messing about with source code, but computers became user-friendly and ordinary people got used to something they could turn on and type a few simple instructions for it to 'just work'. The intricacies of computer code scared rather than fascinated them, and they were happy to pay for the convenience and comfort of buying a brand that they had been persuaded to trust. Companies like Microsoft and Apple were happy to oblige, but there were others who took exception to proprietary licensing, and in particular the agreements that Microsoft was making with PC manufacturers, and were looking to promote alternative business models.

The first version of the Unix Time-Sharing System was completed in 1970 by Ken Thompson, a computer scientist at Bell Laboratories, as a new operating system for what was then cutting-edge computer technology, namely the DEC PDP range of minicomputers. Thompson wrote the original version in assembly language, which limited it to those specific machines. However in 1971 Dennis Ritchie, who worked with Thompson at Bell Laboratories, devised a new programming language which he called C (because it was based on an earlier language called B), and wrote a compiler for it that would run on the DEC PDP-11. By the summer of 1973, Thompson and Ritchie had re-written Unix in the new language, which meant that it could be installed on any computer that could run a C compiler.

The C programming language is beautifully contrived: almost as easy to understand as a high-level language like BASIC or FORTRAN but with low-level structures that allow skilled programmers to write highly efficient programs. It became very popular, with compilers available for a wide range of mainframes and minicomputers. As it was written in C, the Unix operating system followed suit, leading Ritchie to state in 1979:

> It seems certain that much of the success of Unix follows from the readability, modifiability, and portability of its software that in turn follows from its expression in high-level languages. (Ritchie, 1979)

Bell Laboratories was owned by the American Telephone and Telegraph Company (AT&T), which had monopolised the American telephone system for most of the 20th century, entering into a series of agreements with the US government to avoid falling foul of anti-trust legislation. In 1956 it had ended a seven year long anti-trust suit by signing a consent decree which, amongst other things, prohibited it from entering the nascent computer industry. As a result, it could not sell Unix as a commercial product. However the Lab was advised by its lawyers that it could distribute copies for educational purposes, provided it did not advertise or offer backup services (Salus, 1994). Thompson was very happy to share his work, to the extent that he actually read the source code out loud line by line to the West Coast Unix User's Group in 1975. Professor Robert Fabry was responsible for founding the Computer Systems Research Group at the University of California in Berkeley, and attended the readings:

> The first meeting of the West Coast Unix User's Group had about 12 or 15 people. We all sat around in Cory Hall and Ken Thompson read code with us. We went through the kernel line by line in a series of evening meetings; he just explained what everything did ... It was wonderful. (Leonard, 2000)

Legend has it that Thompson distributed copies of the source code to anyone who expressed an interest, accompanying each with a note signed "love, ken" (Raymond, 2003, p. 33).

As a consequence an ad hoc community flourished, offering mutual support and generating various improvements and extensions to Thompson's original work that were in turn distributed in their own right – although not under the name 'Unix' as it seemed likely that AT&T could lay claim to that as a trademark. The most significant was the Berkeley Software Distribution (BSD), which was developed at the University of California in Berkeley by Fabry's Computer Systems Research Group and chosen by ARPA to host the TCP/IP protocols that would eventually underlie the Internet.

Then in 1982, AT&T settled another anti-trust suit that had been dragging on since 1974 by agreeing to end its monopoly of the telephone system. This resulted in the creation of the seven independent 'Baby Bell' telephone companies and the lifting of the prohibitions imposed by the 1956 consent decree, which meant that AT&T was at last able to sell Unix in its own right. Accordingly it established the Unix System Laboratories, which released System V Release 1 (SVR1) as a commercial product in 1983. Two years later, AT&T did indeed take the opportunity to formally register its ownership of the UNIX trademark (Kelly, 1995).

This wasn't the only commercially available version of Unix. Microsoft purchased a licence to use the source code from AT&T in 1979 and announced its own version under the name Xenix the following year, which included elements from the BSD. Microsoft did not sell Xenix directly, but instead sublicensed it to a number of hardware manufacturers, and also to The Santa Cruz Operation (SCO), which in 1983 released a version for the IBM PC. By this time Sun Microsystems, Hewlett-Packard and DEC had also released their own versions of Unix in the form of SunOS, HP/UX and Ultrix.

The commercialisation of Unix jeopardised the collaborative nature of its development and risked fragmenting the market, as companies did what they could to differentiate their versions from those of their competitors. Exactly who had developed which component parts, and who had a licence to use what was becoming increasingly confused, and there was growing tension between those who wanted to keep the source code in the public domain, and those who wanted to capitalise on their intellectual property. In an effort to clarify the issue, the University of California entered into an agreement with AT&T in March 1986 which recognised that the current versions of UNIX and the BSD) each contained sections of code that belonged to the other party, and formally allowing AT&T to distribute software that contained elements of the BSD to other companies. AT&T paid the University $1,750 for the licence and for "one physical copy" of the relevant programs (University of California, 1986).

One person who was very concerned by the actions of AT&T and others, which he felt threatened the collaborative environment in which Unix had thrived, was Richard Stallman, a programmer at the MIT Artificial Intelligence Laboratory. Stallman has been described as "the last true hacker", in that he has remained true to the moral code followed by the programmers and hackers who came together at

MIT and similar organisations through the 1960s and early 1970s, a code that Steven Levy defined as the Hacker Ethic in his 1984 book *Hackers: heroes of the computer revolution.* Underlying the Hacker Ethic is the belief that computers have the potential to change people's lives for the better, and that programming is a creative endeavour that has a beauty of its own. Hackers believe that information should be freely shared in a decentralised fashion, with no controlling authorities or bureaucracies. According to the Hacker Ethic, anyone who so wishes should have free access to the source code of any software program so that they can learn from it and participate in its development in a spirit of collaboration and community.

Stallman's response to the commercialisation of Unix was the GNU Project, which he initiated in 1983 with the aim of developing an operating system compatible with Unix, in that it would be able to run Unix programs, but not containing any proprietary code, so that it could be freely available to anyone who wanted to work with it. As he stated in his original announcement:

> I consider that the golden rule requires that if I like a program I must share it with other people who like it. I cannot in good conscience sign a nondisclosure agreement or a software license agreement. So that I can continue to use computers without violating my principles, I have decided to put together a sufficient body of free software so that I will be able to get along without any software that is not free. (Stallman, 1983)

Recursive acronyms appeal to the hacker mentality: GNU stands for GNU's Not Unix.

Central to Stallman's efforts was the concept of 'free software', although he was keen to emphasise that he was using the term in the sense of 'free speech' rather than 'free beer'. Free software implies the freedom to study and alter how a program works, and to distribute your own modified versions to others, neither of which is possible unless you have access to the source code. Stallman was at pains to point out that free software can also be sold, perhaps in a compiled form with a manual to make it easier to use, but he was determined to ensure that free software could only be distributed if it continued to adhere to these principles. He did not want to follow AT&T or the University of California in allowing others to sell versions under a proprietary licence.

Stallman therefore adopted the principle of 'copyleft', a play on words that effectively reverses the rights bestowed by conventional copyright law. If a work is copyleft then anyone is free to copy and modify it, but the result can only be distributed if it bestows the same rights on others. Stallman crystallised these ideas into the GNU General Public License (GPL), first published in February 1989 by the Free Software Foundation which he established in 1985 to promote the distribution of free software. The GPL specifically allows licensees to distribute copies or modified versions of a program to others, but only if such distributions are also covered by the GPL. Such distributions must acknowledge the copyright of the original, and you can charge for "the physical act of transferring a copy" or for additional services such as a warranty or technical support. However you can't stop your customers from distributing the program, or anything derived from it, free of charge (Free Software Foundation, 2007).

Nowadays most people refer to such software as 'open source' rather than 'free', so as to avoid the implication that it is free of charge. This term was introduced following the announcement by Netscape in January 1998 that it was making the source code of its web browser freely available, which led to the establishment of the Open Source Initiative (OSI) a month later, with Eric Raymond as president. However the OSI supports a number of licences other than the GNU GPL, some of which do not enforce copyleft, which means that the code can be used as the basis for a proprietary product. As a result, Stallman has always insisted that, while free software most definitely qualifies as open source, the opposite is not necessarily true. While he may have a point, such disagreements alienated ordinary users.

For most of those working with Unix, the launch of the IBM PC in 1981 was of little interest as they didn't regard such devices as powerful enough to run a 'grown up' operating system. It was only towards the end of the 1980s, as machines based on 32-bit processors like the Intel 80386 became available, that the Unix world started to take the personal computer seriously. But by then it was too late, as few PC or Apple users were willing to consider anything other than the operating system that was already installed on their machines. The uncertainty surrounding OS/2 did open a brief window of opportunity – after all, there were versions of Unix already capable of running multiple MS-DOS programs concurrently on 386-based PCs – but no one managed to come up with an effective response before Windows 3.0 effectively closed it again. As Raymond put it in his collection of essays published as *The Cathedral & The Bazaar*:

> ... by the early 1990s it was becoming clear that ten years of effort to commercialize proprietary Unix was ending in failure. Unix's promise of cross-

platform portability got lost in bickering among half a dozen proprietary Unix versions. The proprietary Unix players proved so ponderous, so blind, and so inept at marketing that Microsoft was able to grab away a large part of their market with the shockingly inferior technology of its Windows operating system. (Raymond, 1999)

Meanwhile, Stallman's GNU Project had produced a fairly comprehensive set of utility programs, including text editors, code compilers and various user interface components, but progress was slow and the Project had yet to come up with the crucial kernel that would link them to the underlying hardware of the computer.

It was around this time that Linus Torvalds entered Helsinki University to study computer science, where he became familiar with Unix and the C programming language. University attendance in Finland is essentially free, and students are eligible for government subsidies, so Torvalds was able to spend time on his own projects, and in particular on building a version of Unix that would run on his 80386-based PC. His starting point was MINIX, a Unix compatible operating system for PC compatibles that had been created in 1987 by Andrew Tanenbaum of Vrije Universiteit in Amsterdam as a teaching tool. However Torvalds wanted to build something of his own that he could make freely available outside the academic world. In August 1991 he announced his intentions on Usenet:

> Hello everybody out there using minix –
>
> I'm doing a (free) operating system (just a hobby, won't be big and professional like gnu) for 386(486) AT clones. This has been brewing since april, and is starting to get ready. I'd like any feedback on things people like/dislike in minix, as my OS resembles it somewhat (same physical layout of the file-system (due to practical reasons) among other things).
>
> I've currently ported bash(1.08) and gcc(1.40), and things seem to work. This implies that I'll get something practical within a few months, and I'd like to know what features most people would want. Any suggestions are welcome, but I won't promise I'll implement them :-)
>
> Linus
>
> PS. Yes - it's free of any minix code, and it has a multi-threaded fs. It is NOT portable (uses 386 task switching etc), and it probably never will support anything other than AT-harddisks, as that's all I have :-((Torvalds, 1991)

Torvalds was eager to share his work and for others to improve on what he had done, and his announcement sparked a fast-growing community of developers and users. Bash and GCC (mentioned in his announcement) are both GNU utilities, and the combination of the new kernel with utilities from the GNU Project was increasingly recognised as an important new operating system. Torvalds wanted to remove any barriers to its free distribution, so he initially made his kernel available under a licence specifically prohibiting any kind of fee being charged for its distribution. However he soon realised that such an approach prevented distributors from recouping their costs, so he switched to the GNU GPL devised by Stallman, a move that he later described as "definitely the best thing I ever did" (Yamagata, 1997). The resulting operating system became known as Linux in recognition of Torvalds' contribution, although Stallman has always insisted that it should rightly be called GNU/Linux as it includes many components that were developed for GNU. Indeed it could be viewed as the culmination of the GNU Project.

Software programming is an iterative process that hovers uneasily on the borders between engineering, architecture and craftsmanship. You write the code necessary to implement a particular function of the program; you compile and execute it to make sure that it works; you return to the code and correct anything that's causing a problem; you recompile and retest – and so on until you've ironed out all of the bugs. Then you write some more code to implement another function of the program, you make sure that it works properly with what you've already written, and so on, until you've got the whole thing to a state that you are willing to share. Small programs can be written by a single person: Bill Gates, for example, single-handedly wrote most of the BASIC interpreter that he and Paul Allen licensed to Ed Roberts for the Altair 8800. However programs of any size are written by teams, with each member involved in a different aspect of the design, coding or testing of the project. Some of the biggest teams are to be found at Microsoft: Windows XP, for example, contained around 40 million lines of code and was created by a team of some 4,000 people (Maraia, 2005). However most teams are considerably smaller, with commercial companies rarely employing more than 30 developers on a single project, largely because of the cost of doing so.

In the case of a proprietary product such as Microsoft Windows or Office, this process goes on behind closed doors. Only as the program reaches completion does the company release it to its intended users, perhaps in a 'Beta' version so that the team can iron out any last-minute bugs that only become apparent when tens of thousands of people start using it. Finally, once the team is satisfied with the result, version 1.0 of the program is released and end users can start buying licences to use it. Meanwhile, the company has already started work on the next version.

Eric Raymond likened this process to the construction of a cathedral: "carefully crafted by individual wizards or small bands of mages working in splendid isolation, with no beta to be released before its time" (Raymond, 1999). This is the way in which most proprietary software is developed by companies such as Microsoft or Adobe. However it is not the way that open source software such as Linux is put together, particularly not since access to the Internet became widespread. Instead, as Raymond put it:

> No quiet, reverent cathedral-building here – rather, the Linux community seemed to resemble a great babbling bazaar of differing agendas and approaches ... out of which a coherent and stable system could seemingly emerge only by a succession of miracles. (Raymond, 1999)

In the case of Linux, the process started with Torvalds' August 1991 Usenet posting which, although informal in tone, served to stake his claim to the project in the eyes of the hacker community – in other words, to the community of like-minded individuals who enjoy contributing time and effort to such projects. It was an invitation to take part, and in return receive copies of the source code as the project developed, together with recognition of any contributions you might make. No longer is there the strict division between those on the inside who are part of the development team, and those on the outside who will eventually be allowed to pay for the privilege of using the software. Instead, everyone involved takes on the role of coder, tester and user. As project owner, it is understood that Torvalds takes the final decision as to what code actually goes into his project, but the code itself is developed (and is still being developed) through an informal and communal process.

This merging of roles is one of the strengths of open source development. In a closed proprietary process, software is only released to the end user once it is 'finished', and then only in an executable form. Users are invited to report problems, but these tend to be along the lines of, "It crashed when I did this," leaving the development team to try to work out what happened and create some sort of 'patch' that will fix the problem. In an open source project, by contrast, the source code itself is being published to anyone who is interested at frequent intervals right from the start, with new versions sometimes appearing several times in one day. As a result, bug reports are coming back all the time, not only stating that a problem exists but indicating the lines of code that are causing the problem, and often with a solution attached.

However this merging of roles is also a major weakness, as it encourages those involved in the project to think of their relatively small community as their user base, ignoring the wider world of non-hackers who are busy buying computers with operating systems already installed so that they can get on with writing spreadsheets or playing games. To compound the problem, there is a tendency for hackers to treat such people with disdain, which does little to further their cause.

The open and somewhat chaotic manner in which Linux evolves makes it hard to pin down, particularly for non-hackers used to receiving a new version of Microsoft Windows or Mac OS as a discrete package once every few years. By contrast, Linux is an ever-changing and expanding collection of projects that share little except their reliance on Torvalds' kernel. One installation can be very different from another, depending on the hardware on which it runs and the particular tools and supporting programs that have been included to support its intended use. As a result, installing Linux from scratch is a complicated and error-prone process, so most people use a distribution or 'distro', which is a ready-made package containing everything that a particular set of users need to install and work with the operating system. Anyone can release a distro, and it is quite acceptable for companies to charge for its compilation, delivery and support. Furthermore, while most distros are open source, some do include proprietary components, which has caused controversy. The Free Software Foundation lists the distros that conform to its principles on the GNU website, and explains why some of the more popular ones are not included (Free Software Foundation, 2013).

An early distribution was the Softlanding Linux System (SLS), which became available in May 1992, less than a year after Torvalds' original announcement, and included TCP/IP for connection to the Internet and an implementation of the X Window System which provided support for graphical user interfaces (although it was to be a few years before any actually appeared). Soon after came Slackware, which Patrick Volkerding created while studying at Minnesota State University Moorhead in an effort to improve on SLS. The name is a reference to the quality of 'Slack' as promoted by the Church of the SubGenius, a satirical organisation created in the late 1970s and popular amongst students at the time. As Volkerding explained:

> I think I named it 'Slackware' because I didn't want people to take it all that seriously at first … I thought Peter MacDonald (of SLS) would take a look at what I was doing and would fix the problems with SLS. Instead, he claimed distribution rights on the Slackware install scripts since they were derived from ones included in SLS. I … told Peter I wouldn't make other changes to Slackware until I'd written new installation scripts to replace the ones that came from SLS. I wrote the new scripts, and after putting that much work into things I wasn't going to give up. I did everything I could to make Slackware the distribution of choice, integrating new software and upgrades into the release as fast as they came out. It's a lot of work, and sometimes I wonder how long I can go on for. (Hughes, 1994)

Slackware is still available today, and Volkerding has been nominated as a BDFL (Benevolent Dictator for Life), a title bestowed on a small number of project founders in the open source community who are considered worthy enough to have their views taken seriously. Stallman and Torvalds are both acknowledged BDFLs.

Another early distribution came from the Debian Project, which was set up to counter the appearance of a number of "bug-ridden and badly maintained" commercial distributions and "finally create a distribution that lives up to the Linux name." To further quote from The Debian Manifesto:

> It is also an attempt to create a non-commercial distribution that will be able to effectively compete in the commercial market. It will eventually be distributed by The Free Software Foundation on CD-ROM, and The Debian Linux Association will offer the distribution on floppy disk and tape along with printed manuals, technical support and other end-user essentials. All of the above will be available at little more than cost, and the excess will be put toward further development of free software for all users. (Murdock, 1994)

Despite being sponsored by the Free Software Foundation in its early days, Debian is not currently on Stallman's list, primarily because the Project supplies 'non-free' software alongside the distribution itself.

The two most significant commercial distributions came from the German company Software und System Entwicklung (SuSE), with a version based around the Slackware distribution, and from Red Hat in North Carolina, which released Red Hat Commercial Linux a year or so later in October 1994. Both were released as packages which could be purchased for a nominal price ($39.95 in the case of Red Hat). What you got for your money was a manual and a set of CD ROM disks containing not only the necessary executable files but also the full source code. Alternatively, in the spirit of free software, you could download the same files free of charge over the Internet. The SuSE distribution became popular partly because it included YaST (short for Yet another Setup Tool), which made it easier to configure and manage an installation. Similarly the Red Hat distribution introduced the rather more conventionally named Red Hat Package Manager (RPM), which helped you decide what features to include within an installation. Both of these tools were released under the GPL, which meant that, for example, SuSE was able to include RPM in its distribution when it became available.

Commercial releases helped to make Linux acceptable to the business community, which felt more comfortable paying real money for a recognisable brand than downloading a bunch of files created by a largely anonymous group of hackers, even if they were free. As a result, these companies became increasingly successful through the 1990s to the extent that, by 2003, Novell was prepared to pay $210 million for the rights to SuSE Linux, then second only to Red Hat in terms of market share. At the same time, IBM invested $50 million with Novell on the basis that many of its customers were using SuSE Linux on its machines. As for Red Hat, when it floated on the stock market in August 1999, its share price tripled in the first day, resulting in a company valuation of some $3 billion (Shankland, 1999).

For Microsoft, such deals were further evidence of the threat that Linux represented. This was not a good time for the company. Anti-trust proceedings had gone to trial in October 1998, and by 2000 it was facing the prospect of being broken in two by its own government. Windows dominated the desktop but the picture was not so rosy on the server, where Windows NT was coming under serious competition from Linux, particularly when it came to serving up websites.

The first websites were delivered with the help of a small program created at the NCSA at around the same time as Marc Andreessen and Eric Bina were developing the Mosaic web browser (see page 169). Like Mosaic, the program ran under Unix, and by 1995 it had turned into a more sophisticated application managed by a team of developers known as the Apache Group – the name was intended to evoke the native American tribe making a 'last stand' against the establishment, but other team members felt it appropriate because they were indeed building 'a patchy web server'. Early releases were open source but did not support the 'copyleft' principle. However in 2004, the Apache Group moved to a new licence which is accepted by the Free Software Foundation as being properly 'free'.

Linux and Apache were inevitably going to do well on the server side of things because servers tend to be managed by engineers who share many of the traits of the hacker community, and may indeed qualify as members. As a result, Apache soon became the web server of choice, and remains so today, with a recent Netcraft survey revealing that over 52 per cent of websites are served up by Apache Web Server. It is quite possible to compile Apache Web Server to run under Microsoft Windows, however other surveys show that over 65 per cent of web servers run some form of Unix (Q-Success, 2013). As Raymond put it: "The killer app of Linux was undoubtedly the Apache Web Server" (Moore J. , 2001).

However Torvalds had always intended Linux to be an operating system for the desktop as well as the server. As he said in a 1998 interview:

> I was personally never very worried about Linux as a server platform - servers are 'easy' compared to desktops, because in the end servers are fairly anonymous - you only see the network behaviour of a server, while with a desktop system it's much more of a 'complete immersion' environment, not just the network part. So I think that Linux is making more waves in the server area simply because it's the easier market to enter. I still personally consider desktops and personal computing to be the 'final frontier' in some sense. (Heise, 1998)

Reasons to buy into Linux on the desktop were increasing. The first graphical user interfaces appeared with the KDE (Kool Desktop Environment) project initiated by Matthias Etrrich in 1996, and GNOME a year later as part of the GNU project. The Mosaic web browser was already open source, and in 1998, Netscape started the Mozilla project to promote an open source version of Netscape Navigator. GIMP (GNU Image Manipulation Program) came out of a project started at Berkeley in 1995, and was soon recognised as a viable open source alternative to Adobe Photoshop. Sun Microsystems acquired StarOffice in 1999 and released its source code under the GNU GPL a year later as OpenOffice, which it described as "the single largest open-source software contribution in GPL history" (Sun Microsystems, 2000). The release included facilities for reading and writing documents that were compatible with Microsoft Office.

Nevertheless, given Microsoft's proficiency at persuading hardware manufacturers to install Windows, getting Linux on to the desktops of 'ordinary' users – users who weren't interested in Linux for its own sake – was always going to be tough. For a start, many of those open-source applications were also available in versions that ran under Windows. Then there was the choice of graphical user interface that presented the Linux user, a choice alien to those brought up on Windows or the Apple Macintosh desktop, where the user interface is defined by the operating system.

Open source advocates even took their cause to the streets through the Windows Refund movement, launched in January 1999. This was based around a little-known clause in the licensing agreement which you must accept before you can use the copy of Microsoft Windows that comes with a new PC. Most people simply accept the agreement without reading it; however the clause states that anyone declining the agreement should contact the machine's manufacturer to discuss a refund.

The movement declared 15 February to be Windows Refund Day, which involved some 100 supporters marching on Microsoft's Silicon Valley office in Palo Alto holding placards bearing slogans such as "Pro Choice" and "Windows? Thanks, but no thanks!" Unfortunately the offices were closed for the Presidents' Day holiday, but Microsoft nonetheless put up a banner stating "Microsoft Welcomes the Linux Community" and laid on cold drinks for the protesters and the press. They also distributed what became known as the "Dear Valued Customer" letter, which pointed out that protesters should approach the manufacturer of their PC for a refund, and not Microsoft itself. It also emphasised the fact that users did have a choice of operating system – a message aimed as much at the Justice Department as at the protesters. Nevertheless, considerable coverage ensued and the event was considered a success.

The initial response from most PC manufacturers was that refunds would only be given on return of the whole package, including the hardware itself. However in some cases persistence did pay off, with refunds reported to be in the region of $25 to $50.

Just how concerned Microsoft was by these developments became apparent in October 1998, when a couple of confidential internal documents were sent anonymously to Eric Raymond at the Open Source Initiative. They became known as the Halloween documents because they appeared at around that date. The documents had been emailed internally to a number of people, including Bill Gates, Steve Ballmer, Paul Maritz and Jim Allchin, and presented a detailed analysis of open source software, with a particular focus on Linux from both a technical and a business perspective. To quote from the Executive Summary of the first document:

> Open Source Software [OSS] ... poses a direct, short-term revenue and platform threat to Microsoft – particularly in the server space. Additionally, the intrinsic parallelism and free idea exchange in OSS has benefits that are not replicable with our current licensing model and therefore present a long term developer mindshare threat. (Valloppillil, 1998)

Some commentators suggested that the leak may have been deliberate, as it provided evidence to the court that Microsoft did face real competition, even on the desktop. However one section from the documents was particularly damaging because it suggested that Microsoft might "de-commoditize protocols and applications" in an effort to combat the threat. Such a tactic involved extending the accepted standards by which one program works with another, so rendering the existing standard obsolete in the face of Microsoft's new, and possibly better, solution. The strategy became known as 'embrace, extend, extinguish', in reference to a phrase that Microsoft's Paul Maritz is alleged to have used in a 1995 meeting with Intel (USA v. Microsoft, 1998).

Even though still tied up in anti-trust proceedings, Microsoft actually used the tactic when it released the Windows 2000 operating system, complete with an implementation of a well-established security protocol known as Kerberos. Originally developed at MIT during the 1980s, Kerberos defines a process by which machines can identify themselves to each other using 'tickets' issued by a third machine (hence the name, Kerberos being the three-headed dog that guards the entrance to Hades).

As with many protocols, Kerberos allows anyone to extend the protocol in order to support additional features, which Microsoft did. However Microsoft chose not to publish the details of its extension, which meant that while a machine running Windows 2000 could use Kerberos to gain access to other machines, regardless of the operating system they may be running, those machines could not access a Windows 2000 machine in the same way unless they supported Microsoft's extension. As only Microsoft knew how to implement the extension, they too would have to be running Windows.

This information was presented to the court on 28 April 2000 in a paper written by Rebecca Henderson, a professor at MIT, at which point Microsoft pointed out that it had in fact published the details of the extension the day before

(Gates D. , 2000). This was true; however they were contained in an executable file that could only be read on acceptance of a non-disclosure agreement which stated that the details of the extension were a "trade secret" and could only be disclosed to full-time employees of your own organisation if they too agreed to keep them secret, a condition that ensured that the extension could not legally be supported by open source software such as Linux.

Microsoft also adopted the age-old strategy of instilling Fear, Uncertainty and Doubt (FUD) in its customers, a strategy employed to good effect by IBM in the 1970s to prevent its mainframe customers from considering alternatives. In an interview with the *Chicago Sun-Times* in June 2001, Microsoft CEO Steve Ballmer took the opportunity to portray Linux as "a cancer that attaches itself in an intellectual property sense to everything it touches." This was a reference to the 'copyleft' principle enshrined in the GNU General Public License. Ballmer continued, "The way the license is written, if you use any open source software, you have to make the rest of your software open source" (Newbart, 2001). Ballmer's statement was a gross distortion of the actual wording of the licence, but sufficient to engender a degree of FUD in its audience.

The full thrust of Microsoft's attack on the open source movement came some five years later, with the announcement in November 2006 of "a patent agreement covering proprietary and open source products" between Microsoft and Novell, which now owned SUSE Linux (Microsoft, 2006). As Ballmer explained at a technical conference shortly after the announcement was made, Microsoft had entered into the deal because Linux "uses our intellectual property" and he wanted to ensure "the appropriate economic return for our shareholders from our innovation" (Sjouwerman, 2006). By "intellectual property" Ballmer was referring to the many thousands of software patents that Microsoft had claimed through the intervening years.

Patents are only meant to be granted if the product or process described is a genuinely innovative solution to a particular problem: they are not meant to be granted if the solution presented is obvious, or if there is evidence of 'prior art'. However patent lawyers are adept at drafting patents in exceedingly broad and confusing terms, and patent offices are inevitably understaffed. Disputing the validity of a patent is a costly and lengthy process, so it has become common for corporations to file a large number, well aware that many would not survive a challenge. However they are also aware that the same is true of their competitors, so the result is a stand-off which has the added benefit of preventing newcomers from entering the market. It also opens up the possibility of cross-licensing deals, such as that entered into by Microsoft and Apple in 1997 (see page 207).

The nature of the GNU General Public License makes cross-licensing impossible, so instead each company agreed to make a payment to the other in exchange for a release from any potential liability that its customers might hypothetically have incurred through patent infringement. As Microsoft sold rather more copies of Windows than Novell did of Linux, the net result was a payment from Microsoft to Novell of a little over $100 million. In addition, Microsoft agreed to purchase coupons from Novell that it would distribute to its customers, entitling them to a one-year subscription to SUSE Linux Enterprise Server, bringing with it the assurance that they were buying a version that was immune from prosecution with regards to any patents that it might infringe. This part of the deal is estimated to have cost Microsoft another $240 million (Parloff, 2007).

This was a convoluted arrangement indeed, but the end result was exactly what Microsoft wanted. Microsoft claimed that 235 of its patents were being infringed by open source software, including 42 by the Linux kernel and another 65 by the graphical user interfaces included with most distributions. The company was careful not to specify exactly which patents, as that would make it easier for hackers to challenge their validity in court. Such challenges might well have resulted in many of the patents eventually being dismissed after lengthy court proceedings, but until that happened, Linux users faced the possibility of being sued. Furthermore, Microsoft did not actually want to go down that road, as many of these users were large corporations who were also valuable Microsoft customers. By entering into this deal, Microsoft was able to offer such customers a 'safe' copy of Linux and so sustain the notion that Linux itself violated its patents – a state of affairs that Microsoft felt well worth paying for. As Ballmer himself put it:

> Novell pays us some money for the right to tell customers that anybody who uses SUSE Linux is appropriately covered. [This] is important to us, because [otherwise] we believe every Linux customer basically has an undisclosed balance-sheet liability. (Sjouwerman, 2006)

Mark Webbink, a representative of Red Hat, summed it up rather differently:

> It allowed [Microsoft] to go out and trumpet that, "see, we told you Linux infringed, and these guys are now admitting it." (Parloff, 2007)

Nevertheless the deal was taken up by many of Microsoft's larger customers, including Credit Suisse, Deutsche Bank, HSBC and Wal-Mart, in an effort to protect their Linux installations.

Despite such antics, many within Microsoft did recognise the benefits of the open source process – after all, many of its own employees could themselves be described as hackers. Furthermore, Microsoft had actually been licensing Windows source code to universities and the like for educational and research purposes long before the Halloween documents came to light, and more recently to select corporations within the United States and leading hardware manufacturers to help them to build more compatible products. In a statement released in 2001 under the title 'Accessibility with Responsibility', senior vice president Craig Mundie outlined the company's Shared Source Initiative, but was careful to distance it from the "viral nature" of the GNU GPL:

> Microsoft's shared-source approach – like the open-source model – provides opportunities for researchers, customers and outside developers to examine Microsoft source code and help the company improve it. However ... maintaining control over its source code is crucial – not only for Microsoft's long-term profitability but also for preserving the stability, compatibility and security of its software for customers. (Microsoft, 2001)

Microsoft released the source code to such people under a number of different licences, and in July 2007 announced that it had submitted two of them, namely the Microsoft Public License (Ms-PL) and the Microsoft Reciprocal License (Ms-RL), to the Open Source Initiative for approval. Three months later, they were both approved. Even the Free Software Foundation recognises them as being 'free', although it advises against their use because they are not compatible with the GNU GPL.

In September 2006, partly in response to obligations arising from the anti-trust proceedings, the company published the Microsoft Open Specification Promise. This was a statement to the effect that it would not assert any legal rights that might arise from any patents that it held with respect to a specific range of technologies, including the Kerberos extensions that had caused such a ruckus earlier. The Promise was broadly welcomed by the open source community but with reservations, as it did not necessarily extend to future derivations of the listed technologies, and Microsoft declined to comment on "how our language relates" to the terms of the GPL.

Microsoft still had problems with the copyleft principle enshrined in the GNU GPL. However in July 2009, the company actually contributed 22,000 lines of code to the Linux kernel itself, code that made it easier for a Linux distribution to work with a machine running Windows Server 2008. The inclusion had obvious benefits to Microsoft itself, but could only be accepted under the terms of the GPL. Nevertheless, as project leader Greg Kroah-Hartman explained, "They abided by every single rule and letter of what we require to submit code. If I was to refuse this code it would be wrong." For Jim Zemlin, executive director of the Linux Foundation, it represented a significant turning point: "Obviously we are tickled about it. Hell has frozen over, the seas have parted!" (Fontana, 2009).

16.   Catching the wave

At the time of writing, articles have been appearing in the press heralding "The Death of the PC" and "PC sales continue to collapse". Such declarations are based on reports that sales of desktop computers have been steadily declining since 2010, and are projected to continue to do so as more people turn to tablets and other mobile devices. They are also fuelled by Steve Jobs' comments, made around the launch of the iPad, when asked whether tablets would replace PCs:

> When we were an agrarian nation, all cars were trucks, because that's what you needed on the farm. But as vehicles started to be used in the urban centres, cars got more popular. Innovations like automatic transmission and power steering and things that you didn't care about in a truck as much started to become paramount in cars. PCs are going to be like trucks. They're still going to be around, they're still going to have a lot of value, but they're going to be used by one out of x people. (Jobs, 2010)

It was obviously in Apple's interest to stir such speculation, but Jobs did have a point. Elsewhere he talked about a 'post-PC era', in which the role of the traditional desktop can indeed be compared to that of the modern-day truck, used only for specialised tasks and abandoned once we can return to more responsive and convenient devices.

However, it is a little early to write off the desktop completely: the basic setup of sitting at a desk in front of a screen, keyboard and mouse is likely to remain useful for some time yet. After all, it's a natural working arrangement that we've been using ever since scribes started propping up manuscripts in front of them so they could copy the contents onto parchments laid out on the desks at which they sat. The walk-in three-dimensional display used by Michael Douglas in the 1994 film *Disclosure*, or the hand-waving projections controlled by Tom Cruise in the 2002 film *Minority Report*, do add dramatic effect, but they are hard work compared with the natural and relaxed movement of mouse on desktop. As for the keyboard, it has long been suggested that we should be talking to our computers by now, but few of us actually do this, even though accurate speech recognition has been with us for some time.

What has changed is that the desktop computer has been augmented by new devices such as the tablet or the smartphone, which allow us to leave our desks and continue computing while walking down the street or sitting in a café. In a 1991 article for *Scientific American*, Mark Weiser of Xerox PARC suggested that we are entering a time of 'ubiquitous computing', which he went on to describe as the Third Wave:

> The First Wave was many people per computer. The Second Wave was one person per computer. The Third Wave is many computers per person. (Lewis, 1994)

As a consequence, the desktop computer no longer represents the forefront of technology. The spotlight has moved on, and in that sense its story has indeed come to an end.

However it's a story that reveals much about the way in which business actually works within a modern economy. For a start, look at the various roles that central government has played through its unfolding. On the one hand are the anti-trust actions instigated against Microsoft by the Department of Justice in the late 1990s, largely at the behest of Microsoft's competitors, which were an unmitigated and extremely expensive disaster that had little effect on their target. Indeed they proved to be against the interests of the ordinary user, in preventing Microsoft from delivering the seamless experience that Apple, which had complete control over its operating system, was able to provide.

On the other hand, in the absence of any real commercial demand, it is unlikely that companies like Texas Instruments, Fairchild or Intel would have developed so fast without large-scale government projects such as the Apollo moon landings or the nuclear defence programme. Driven by Cold War paranoia, and stung by the Soviet Union's early advances into space, the Advanced Research Projects Agency (ARPA) funnelled unprecedented amounts of government money through universities and research institutions, often with little supervision. Advances in user interface design, communications and much else came out of projects instigated at Berkeley, Harvard, MIT and Stanford, and led eventually to the Apple Macintosh, Windows and the World Wide Web. The Internet, in particular, would not have grown into the homogenous and resilient global network that it became without the sensitivity that ARPA showed in handing it over to the private sector. Furthermore, these institutions were generally happy to let their students benefit fully from their work, even if the results were multi-million dollar businesses. It's a shame that many of these selfsame businesses now see fit to avoid paying the taxes that make such generosity possible.

It has become fashionable to sing the praises of the free market, but the public and non-profit sectors do have an important part to play in providing services that are not suited to the dynamics of the private sector, where each party is busy protecting its own interests. Reflecting on the development of the Web, Tim Berners-Lee said:

> The Web should be a universal space. Should be usable for any sort of information ... In order to allow that, one of the things which it should not be is proprietary. There shouldn't be money payable to one particular company for using the Web, otherwise there would be a single point of control and a single point of failure, for one thing. And also there would be people who just didn't want to do that, and so there would be a whole lot of information which was excluded from the Web as a result. So the openness was really a part of the universality ... All the things which happened on the Internet ... were produced in a non-commercial way, with the vision of people communicating better. (Berners-Lee, 2001)

This story also serves to demonstrate the shortcomings of simplistic free-market models. The business of manufacturing and selling computer hardware bears comparison to that of the car or the TV, in that each individual unit costs a considerable amount of money to create and distribute. As the technologies involved coalesced around a standard, namely the Intel-based PC, the opportunities for one manufacturer to distinguish their products from another diminished, and they were forced to rely increasingly on brand awareness. The customer has an understanding of what it means when a device bears an Apple logo, and how that differs from a Dell or a Hewlett-Packard, even though the underlying hardware may be much the same. There are economies of scale, but competition is fierce, and no one brand can dominate the market for any appreciable length of time. Apple only succeeded in doing so by establishing itself firmly in that 'high-right' corner favoured by Gassée, as a supplier of luxury items at premium prices.

Running alongside this are two industries, namely that concerned with microprocessors and that concerned with software, which function very differently, in that they benefit from the economies of scale to an unprecedented degree. The cost of developing a new microprocessor or a new software program is enormous –

particularly if you're talking about something like an Intel Core i7, with well over a billion transistors crammed on to a sliver of silicon less than 2cm across, or a modern operating system that can involve 100 million lines of code. Against that, the cost of creating copies, even millions of copies, is miniscule – and the more copies you create, the smaller the proportion of that initial investment each copy has to bear.

The larger the company, the more that it can benefit from such economies and the harder it becomes for others to compete. In the desktop market, Intel's only real competitor is Advanced Micro Devices (AMD), which has been producing Intel compatible microprocessors since 1982 when IBM demanded a second source for the Intel 8088. However it is Intel that has defined the standard. Microsoft's dominance of not only desktop operating systems but also word processing and other office applications was secured in 1990 with the success of Windows 3.0 and Microsoft Office 1.0. Only Apple offered any real alternative, largely thanks to the quality of its products and its occupation of niche markets such as graphics and publishing towards the end of the 1980s, before Windows became a viable alternative. According to some surveys, one billion people – a seventh of the global population – use Microsoft Office either at home or at work (Graham, 2013).

Economists and politicians traditionally regard such monopolies as undesirable. However most of those involved in desktop computing, including customers and software companies, have benefited hugely from the dominance of Microsoft and Intel and the 'Wintel' platform that they created. Software companies can release Windows applications confident that they will work on most desktop computers around the world. Customers can transfer documents knowing that most of their friends and colleagues can open and work with them. Furthermore, Microsoft has generally done what it can to ensure that each successive version of its operating system is backwards compatible, so that users haven't had to replace their applications when they upgrade. Anyone who can remember the days of incompatible operating systems and disk formats will appreciate these benefits. It is only companies like Digital Research, Netscape or Oracle, which were looking to establish alternatives, that lost out.

The problem is that these two industries are natural monopolies, and attempts to curtail the companies involved inevitably end in disaster. Furthermore, examples are becoming more common as the Internet matures, delivering new economies of scale that have allowed the likes of Amazon, Facebook and Google to build platforms that span the globe. This is not a treatise on economics or corporate law, but I would suggest that governments recognise this phenomenon and do what they can to encourage such companies to acknowledge the position that they are in and the responsibilities that it brings.

Many of the tactics adopted by Microsoft in an effort to secure its control of the PC platform can at best be described as ruthless and at worse as immoral or even downright evil. Certainly, the company resorted to spreading Fear, Uncertainty and Doubt (FUD) when it felt threatened, or a policy of Embrace, Extend and Extinguish when something came along that did not fit with its plans. However criticism does beg the question as to what we expect of a company operating within a capitalist economy. In the early days there was a hope, particularly amongst the West Coast hacking community, that this could become an industry based on values other than cut-throat competition and profit maximisation. But in the end, as the money started rolling in, it has turned out to be more of the same. As Gary Kildall's son recalled, some years after his father's premature death following a blow to the head: "My father was devastated when he discovered that the computer industry was just as ruthless as any other" (Kildall, 1997).

On the other hand it can be argued that Microsoft was simply doing what was necessary to maximise shareholder value, which is, after all, what companies are supposed to do in a capitalist economy. Only when companies act ruthlessly to secure their position and disadvantage their competitors can market forces ensure the survival of the fittest and the best use of the resources available. On this basis, it is up to governments to legislate against undesirable behaviour – companies should not be expected to concern themselves with such matters.

Certainly an abiding impression from the anti-trust proceedings is that Gates did not accept that he had done anything wrong; he was simply upset that his competitors had seen fit to move the battle from the marketplace to the courtroom. Nevertheless, one positive result was some degree of acceptance that, as guardian of the Windows platform, Microsoft had a responsibility not only to its customers but also to other companies that operate within the Windows 'ecosystem'. Whether such a painful and expensive process was necessary to achieve such a result is open to question.

Although many other companies have been involved, the struggle between Microsoft and Apple, and the platforms they created, does dominate this story.

237

Indeed much of this history has been shaped by the reaction of four exceptional people to the MITS Altair 8800 when they first encountered it back in 1975. Bill Gates and Paul Allen were by nature programmers, so their first instinct was to develop software for this new machine, and then for other microcomputers as they appeared. When, a few years later, IBM took the unprecedented decision to enter the market with a machine built almost entirely from off-the-shelf components, Microsoft was there to deliver the all-important operating system – not just to IBM but to all of the clone manufacturers that went on to hijack what it meant to be 'PC compatible'. Since then, Microsoft has largely concentrated on delivering the software that defines and maintains the industry's most widely adopted computing platform.

Steve Wozniak was a talented programmer, but his main interest lay in assembling integrated circuits into elegant and efficient devices. His immediate reaction on encountering the Altair 8800, just a few months after Gates, was that he could build something better, which turned out to be the Apple 1. Steve Jobs was not a programmer, but he did have a consuming vision of what it could mean for a computer to be truly 'personal', and as a consequence was able to grasp the significance of what Xerox had at PARC in a way that Xerox itself singularly failed to do. His response was to take tight control of both the hardware and the software and turn it into something that fully encapsulated the elegance and simplicity that would make his devices desirable and fun. In the process, he established Apple as one of the strongest and most distinctive brands the world has ever known.

Of course the story would have been different had Gary Kildall had followed Bill Gates' advice and signed up to produce a 16-bit version of CP/M for the IBM PC. The result would have strengthened the CP/M platform, leaving little reason for Microsoft to develop MS-DOS. Instead Gates would have concentrated on developing applications, both for CP/M-86 and for the Apple Macintosh when that came along. Digital Research would have responded with the GEM graphical user interface, but without Microsoft's leverage on the applications side, would have had little to counter Apple's 1985 legal challenge. Depending on how the relationship between Gates and Kildall developed, Gates may still have tried to get Apple to license the Macintosh user interface to other manufacturers, but Gassée's objections would still have been valid. Meanwhile VisiCorp would be working on Visi On, and Microsoft would start thinking about building a windowing environment for its own applications. Initially Interface Manager would have run as a CP/M application, but Microsoft would inevitably have started adding lower-level functionality to the program, and the result might well have ended up looking much like Windows 3.0.

In the event, by the mid-1980s Microsoft had secured its position as sole supplier of the MS-DOS operating system which played a large part in defining what it meant to be 'PC compatible'. The uncertainty surrounding OS/2 did open a brief window into which a version of Unix could have stepped, but Linux was still several years away and the rest of the Unix world too inward-looking to come up with a coherent strategy, before the window was decisively closed by the runaway success of Windows 3.0. By the time that Linux did become a viable alternative, Windows had secured too many users, and supported too many applications, for it to stand a chance.

It has long been understood that technological advances have a disruptive effect on business, changing the nature of the game and destroying the companies that fail to adapt. As Jean-Louis Gassée put it:

> There are waves that are independent of companies, and then the psychology of people makes some companies ride the wave very well, and the psychology of other companies – the executives – causes some companies to miss the wave. (Gassée, 2012)

Or, to paraphrase an old adage, when the internal combustion engine came along, the buggy makers went bust. Our story started with the development of the microprocessor which made possible the vision outlined by Alan Kay in his 1972 paper 'A Personal Computer for Children of All Ages', but only if the industry could produce something that was cheap, reliable and easy to use. Only when a computer cost around the same as TV or hi-fi system would it be able to move out of the back room and into the homes and onto the desks of ordinary people. This was what Apple, Commodore and Tandy set out to achieve, closely followed by the likes of Atari, Sinclair and Acorn.

Of the previous generation, only IBM and Xerox would leave a lasting impression. The IBM PC was extremely successful, not only because of the brand but also because it was a good product at an affordable price. IBM's decision to build it from readily accessible components, and to allow Microsoft free range with its operating system, ensured that it established an industry standard but, as IBM discovered, this was not a standard that IBM would control. Microsoft "rode the bear" (as vice-president Steve Ballmer put it) until it made sense to jump off and instead concentrate on ensuring that MS-DOS remained central to what it meant to be 'PC compatible'. Unable to come up with a viable strategy in the face of competition from clone manufacturers unencumbered by outdated considerations, IBM embarked on an effort to integrate its personal computer range into a largely irrelevant corporate strategy, and in the end only survived by returning to its roots as a supplier of large-scale business solutions.

Of course it's just as easy to miss a wave by jumping too soon as it is by waiting too late. By the latter half of the 1970s, researchers at Xerox PARC were already sitting in front of networked computers that were controlled by mice and displayed icons on the screen. However the technologies and the economics that would make such systems affordable were nearly a decade away. Rather than 'riding the wave' themselves, a group of executives within Xerox opted to show Apple what they had achieved in exchange for an opportunity to invest. Disillusioned, many who had worked at Xerox PARC left to take up positions at Apple and Microsoft, bringing with them the skills and experience that would lead to both the Macintosh and Windows.

The next major wave was heralded by the World Wide Web, making possible a whole new range of applications that gave us all new things to do with our computers. Both Apple and Microsoft were caught on the hop, but it was Microsoft that felt most threatened when Andreessen started talking about an Internet Operating System that would give the user access to computing power that would reside out there in the Internet, without the need for anything as cumbersome as Windows. Microsoft's response put them in court, but the fact remains that Netscape had also jumped too soon. What Andreessen was describing bears comparison with what we have now, namely relatively low-powered smartphones and tablets communicating at high speed with data centres out in the 'cloud' (as we now call it). However at the time, back in the mid-1990s, the Internet was simply not fast enough, reliable enough or ubiquitous enough to sustain such a vision.

And it's those smartphones and tablets that heralded the wave that marks the end of this story, and saw Alan Kay's vision finally come to fruition. Once again, there are some that jumped too soon, although this time they did come back for another try. Despite its many fans, the Apple Newton was far too expensive and nowhere near reliable enough for a wider audience. Microsoft's Pocket PC and Windows Mobile devices appeared some seven years later and offered many of the features that we've come to associate with smartphones. They could collect email and display photographs, for example, and browse the Web. They even introduced a mobile version of Microsoft Office called Pocket Office. However the user interface was not dissimilar to the desktop version of Windows, with few concessions to the mobile format.

Windows Mobile actually did quite well, particularly with business users, but that all changed when Apple released the iPhone in June 2007. Claims that Apple had invented the smartphone did annoy Windows Mobile users, but what Apple had done was truly innovative, creating an operating system fine-tuned for touchscreen mobiles and introducing a marketplace for 'apps' (as applications became known) based on the same business model as its highly successful iTunes Store. The starting price of $499 was not cheap, but Apple's reputation was for delivering premium products at premium prices, so it sold like hot cakes.

Three years later, Apple repeated the trick with the iPad, which is about as close to Alan Kay's Dynabook as you can get, and in 2010 Microsoft re-entered the fray with Windows Phone, a complete rethink of its earlier strategy. However figures for the final quarter of 2012 show Microsoft accounting for less than 3 per cent of worldwide sales and Apple for just 21 per cent, with the dominant share going to Android, which can claim over 70 per cent of sales (IDC, 2013). Android is an open source operating system based on Linux but specifically designed for mobile devices, and has been owned by Google since 2005. It is used by a number of smartphone and tablet manufacturers, the most significant being Samsung. There are certainly parallels to be drawn between the rise of Android on mobile devices and the story of Windows on the desktop, but that's another story.

Timeline

Significant events in chronological order.

1964
Apr: IBM launches the System/360 mainframe and celebrates its 50th birthday.
Jun: BASIC computer language is developed at Dartmouth College.

1965
Mar: Ted Nelson describes hypertext as a component of the Xanadu project.
Mar: DEC launches the PDP-8 minicomputer, starting at $16,200.
Apr: *Electronics* magazine publishes article describing Moore's Law.

1966
Nov: The RESISTORS computer club formed in New Jersey.

1967
Ludwig Braun starts the Huntington Project using BASIC.
Nicholas Negroponte sets up the Architecture Machine Group at MIT.
Feb: First manned Apollo mission ends in disaster.

1968
Apr: The film *2001: A Space Odyssey* is released.
Jul: Robert Noyce and Gordon Moore found Intel Corporation.
Aug: The exhibition *Cybernetic Serendipity: the computer and the arts* opens in London.
Oct: First successful Apollo mission and first live broadcast from American spacecraft.
Dec: Douglas Engelbart conducts the 'Mother of All Demos' in San Francisco.

1969
Assembly language version of UNIX operating system written at AT&T Bell Labs.
Jan: Johnson administration opens anti-trust proceedings against IBM.
Jul: Apollo 11 lands men on the moon.
Aug: Ted Hoff at Intel proposes construction of a 4-bit microprocessor.
Oct: ARPANET becomes active.

1970
Ken Thompson writes Unix Time-Sharing System for PDP-11 at Bell Laboratories.

Jan: DEC announces the PDP-11 minicomputer, starting price $10,800.

Jul: Xerox PARC opens in Stanford Research Park, Palo Alto.

Oct: Intel introduces 1103 memory chip, storing 1,024 bits.

1971

Aug: Gary Boone at Texas Instruments files patent for 'Computer Systems CPU'.

Nov: Intel launches 4004 microprocessor as part of the MCS-4 microcomputer system.

1972

Smalltalk-72 programming language developed at Xerox PARC.

Community Memory set up in San Francisco as an electronic message board.

Apr: Intel introduces 500KHz 8008 microprocessor as part of the MCS-8 family.

Aug: Alan Kay presents his paper 'A Personal Computer for Children of All Ages' in Boston.

Oct: Launch of the People's Computer Company.

Dec: Apollo 17 marks last manned spaceflight beyond Earth orbit.

Dec: First installation of arcade game *Pong* at Andy Capp's Tavern in Sunnyvale, California.

1973

Thompson and Ritchie rewrite Unix using the C programming language.

Intel introduces the Intellec 8 microcomputer at $2,400.

IBM introduces the eight-inch floppy disk drive and the 'Winchester' hard disk drive.

ARPANET connects 30 institutions.

Feb: Micral N microcomputer launched in France by R2E at 8,500 Francs.

Apr: First Xerox Alto prototypes built at Xerox PARC.

1974

Gary Kildall writes CP/M operating system.

First Ethernet network established at Xerox PARC.

First laser printer developed at Xerox PARC.

Mar: Intel releases 2MHz 8080 microprocessor.

Aug: Motorola releases 6800 microprocessor.

Aug: Ted Nelson publishes *Computer Lib*.

Sep: Launch of *Creative Computing Magazine*.

1975

Home *Pong* console goes on sale through Sears and Roebuck.

Jan: MITS launches the Altair 8800 microcomputer in *Popular Electronics*.

Mar: First meeting of the Homebrew Computer Club.

Apr: US evacuation of Saigon, marking the end of the Vietnam War.

Apr: Micro-Soft is founded by Bill Gates and Paul Allen in Albuquerque.

Jun: MOS Technology announces 6502 microprocessor.

Jun: National Enterprise Board set up by the British government.

Jul: The Computer Store opens in Los Angeles.

Aug: Telenet starts operation as first public packet-switching network.

Sep: First issue of *Byte* magazine.

Sep: MOS Technology launches the 6502 microprocessor at $25.

Sep: IBM 5100 Portable Computer launched at starting price of $8,975.

1976

Colossal Cave adventure game written by Will Crowther.

Jan: First issue of *Dr. Dobb's Journal* published.

Feb: Bill Gates writes 'An Open Letter to Hobbyists'.

Mar: First World Altair Computer Convention held in Albuquerque.

Mar: Wang Laboratories introduces the 2200 PCS at $5,400.

Apr: Steve Jobs and Steve Wozniak form the Apple Computer Company.

Apr: Gary Kildall founds Intergalactic Digital Research to sell CP/M operating system.

Jul: Zilog releases 2.5MHz Z80 microprocessor.

Jul: The Apple-1 goes on sale at the Byte Shop in California for $666.66.

Aug: Personal Computing '76 show held in Atlantic City.

Sep: Shugart introduces the 5.25-inch SA400 'minifloppy' disk drive.

1977

Jan: Commodore PET announced at Winter CES in Chicago.

Apr: Apple II launched at First West Coast Computer Faire priced from $1,298.

Jun: Xerox launches the Xerox 9700 laser printer.

Aug: Radio Shack TRS-80 announced at starting price of $600 with display.

Oct: Commodore PET becomes available at starting price of $495.

Oct: Atari Video Computer System (VCS) goes on sale for $199.

Nov: Future Day held at Xerox World Conference.

Dec: Nascom 1 single-board computer goes on sale.

1978

Multi-User Dungeon (MUD) set up at Essex University.

Feb: CBBS, the first computer-based bulletin board, goes live in Chicago.

Mar: BBC broadcasts *Horizon: Now The Chips Are Down* predicting mass unemployment.

May: Intel introduces 4.77MHz 8086 microprocessor.

Jun: *Space Invaders* games console released by Taito Corporation.

Jun: Science of Cambridge MK14 kit becomes available for £39.95.

Jul: Apple Disk II becomes available at introductory price of $495.

1979

Jan: Microsoft moves from Albuquerque to Seattle.

Mar: Acorn Microcomputer released by Curry and Hauser for £80.

Mar: British Telecom launches the Prestel service.

Jun: Intel 8088 microprocessor released.

Jun: Software Arts demonstrates VisiCalc at the National Computer Conference in New York.

Jun: MicroPro launches the WordStar word processing application.

Jun: The Source is announced at Comdex.

Jun: Bell Labs releases Version 7 of UNIX. Microsoft buys a licence to the source code.

Sep: Motorola releases 68000 microprocessor.

Oct: ITV broadcasts *The Mighty Micro* as a six-part series.

Nov: Atari 400 and 800 microcomputers become available.

Dec: Steve Jobs and colleagues visit Xerox PARC.

1980

Jan: Science of Cambridge launches ZX80 for £99.95.

Jan: The Usenet distributed discussion system goes live.

Jan: *A2-FS1 Flight Simulator* released for the Apple II.

Mar: Acorn Atom released for £170 fully assembled.

May: Seagate ST-506 hard disk becomes available.

May: *Pac-Man* games console released in Japan.

Jun: Commodore announces VIC-20 at $299.

Jun: *Zork* adventure game from Infocom available for the TRS-80.

Aug: IBM visits Microsoft and Digital Research to discuss software for IBM PC.

Aug: Microsoft launches Xenix operating system.

Sep: Seattle Computer Products ships 86-DOS.

Oct: Microsoft signs deal with IBM to produce software for the forthcoming PC.

Oct: Epson introduces the MX-80 dot matrix printer.

1981

ARPANET connects 200 computers.

Jan: Commodore VIC-20 launched.

Feb: Charles Simonyi leaves Xerox to join Microsoft.

Mar: Sinclair Research launches the ZX81 for £69.95.

Apr: Xerox Star introduced with prices starting at $16,595.

Apr: SuperCalc spreadsheet program launched at West Coast Computer Faire.

Jun: Sony introduces the 3.5-inch Micro Floppy Disk (MFD) drive.

Aug: IBM PC launched in the US with a starting price of $1,565.

Oct: ACT launches Sirius 1 at SYSTEMS 81 in Munich.

Dec: Victor 9000 launched at COMDEX in Las Vegas.

1982

Jan: US President Ronald Reagan dismisses the anti-trust suit opened in 1969 against IBM.

Jan: Intel announces the 80186 microprocessor.

Jan: BBC broadcasts *The Computer Programme* and launches BBC Microcomputer.

Jan: Digital Research CP/M-86 operating system becomes available.

Jan: Microsoft agrees to produce applications for the Apple Macintosh.

Feb: Intel announces the 80286 microprocessor.

Mar: Telecom Gold launched by British Telecom as a commercial email service.

April: Sinclair launches ZX Spectrum at starting price of £125.

Jun: Columbia Data Products announced the first IBM PC clone, the MPC.

Jul: Science fiction film *Tron* goes on general release.

Aug: Commodore 64 released.

Oct: 3COM introduces Ethernet network cards for IBM PC and compatibles.

Nov: Compaq Portable announced.

Nov: AT&T's Unix System Group releases System III.

1983

Jan: ARPANET switches to the Internet Protocol Suite (TCP/IP).

Jan: Lotus 1-2-3 spreadsheet program launched at $495.

Jan: Worldwide launch of the IBM PC.

Jan: Compaq Portable launched at $3,590.

Jan: Apple Lisa becomes available at $9,995.

Jan: AT&T releases System V Release 1 (SVR1) as a commercial version of UNIX.

Jan: Novell releases NetWare operating system.

Mar: IBM PC XT launched with 10MB hard disk drive at $4,995.

May: The film *WarGames* is released.

May: Microsoft releases Multi-Tool Word and the Microsoft Mouse.

Jun: ACT launches the Apricot PC.

Sep: Teenagers Austin and Poulsen arrested for 'malicious access'.

Sep: Santa Cruz Operation releases Xenix for the IBM PC.

Sep: Richard Stallman announces the GNU Project.

Oct: Microsoft Word becomes available.

Nov: IBM PCjr launched with starting price of $699.

Nov: First demonstration of Microsoft Windows to the press.

Dec: VisiCorp releases Visi On user interface with three applications.

1984

Jan: Apple Macintosh launched for $2,495.

Mar: Digital Research releases Concurrent DOS 3.1 to hardware manufacturers.

Apr: CompuServe announces its Electronic Mall.

May: Phoenix PC ROM BIOS made available for $25.

May: Digital Research demonstrates GEM user interface at Comdex.

May: Hewlett-Packard introduces the LaserJet printer.

Jun: Amstrad CPC launched.

Jun: FidoNet released by Tom Jennings.

Aug: IBM PC AT launched for $5,795 with 20MB hard drive.

Aug: IBM announces TopView user interface at $149.

Sep: Acornsoft releases *Elite* space trading game for the BBC Micro.

Sep: Apple Macintosh 512K 'Fat Mac' launched at $3,195.

Oct: Borland Sidekick 1.0 released.

Oct: Oracle database released for IBM PC XT and AT.

Oct: Prince Philip's Prestel mailbox is hacked.

1985

Jan: Apple LaserWriter announced alongside Aldus PageMaker.

Jan: Atari 520ST announced at Winter CES in Las Vegas for $800.

Feb: Digital Research releases GEM user interface.

Mar: Bill Atkinson starts work on HyperCard application at Apple.

Apr: Compaq Deskpro 286 launched.

Apr: Oracle 5.0 introduced with support for client-server operation.

Jun: Atari 520ST becomes available, running GEM user interface.

Jun: Bill Gates suggests licensing strategy to Sculley and Gassée at Apple.

Jun: AT&T applies to register the UNIX trademark.

Jul: Commodore Amiga 1000 launched at the CES in Chicago for $1,295.

Aug: Microsoft and IBM sign Joint Development Agreement.

Sep: Steve Jobs resigns from Apple.

Oct: Intel 80386 processor announced, with 16MHz clock.

Oct: Richard Stallman starts the Free Software Foundation.

Nov: Microsoft releases Windows 1.0 for $99.

Nov: Apple settles with Microsoft over use of Macintosh features in Windows 1.0.

Dec: The Internet of TCP/IP networks connects 2,000 host computers.

1986

Jan: Apple Macintosh Plus introduced at AppleWorld in San Francisco at $2,599.

Jan: Steve Jobs helps finance Pixar Animation Studios.

Mar: Digital Research releases GEM 2.0 in response to Apple suit.

Jul: The Great Renaming of Usenet groups starts.

Sep: Amstrad PC1512 launched at starting price of £399.

Oct: Compaq DeskPro 386 available at $6,499.

Oct: Computer Fraud and Abuse Act becomes law in the US.

Nov: Seattle Computer's suit against Microsoft goes to trial.

1987

The Internet connects 30,000 host computers.

Mar: Apple Macintosh SE becomes available from $2,899.

Mar: Aldus PageMaker released for PC compatibles.

Apr: IBM PS/2 range launched, starting at $1,695 plus $250 for display monitor.

May: Apple Macintosh II launched priced from $5,000.

Oct: Motorola launches 68030 processor.

Oct: Mac OS System 5 introduced with MultiFinder.

Nov: OS/2 SE 1.0 operating system available from $325.

Nov: Microsoft Windows 2.0 becomes available.

1988

Mar: Apple files copyright suit against Microsoft concerning Windows 2.03.

Apr: Mac OS System 6 released with cooperative multitasking.

May: Microsoft Windows 2.1 released in Windows/286 and Windows/386 versions.

May: Digital Research release DR DOS 3.31.

Jul: OS/2 EE 1.0 released at $795.

Sep: 'Gang of Nine' publish the specification for the EISA expansion bus.

Oct: NeXT Computer is demonstrated to the press priced from $6,500.

Nov: OS/2 SE 1.1 released with Presentation Manager.

1989

Feb: The GNU General License (GPL) published by the Free Software Foundation.

Mar: Tim Berners-Lee submits proposal for WorldWideWeb, which is ignored.

Apr: Intel launches the 80486 running at 20 and 25MHz.

Aug: Microsoft Office 1 for Macintosh released.

Sep: Macintosh Portable launched at starting price of $5,799.

Nov: Compaq launches first EISA machines in DeskPro 486 and SystemPro.

Nov: Microsoft Word for Windows 1.0.

Nov: The World is first commercial provider of dial-up Internet access in the US.

1990

Feb: ARPANET is closed down. Internet connects nearly 300,000 computers.

May: Digital Research release DR DOS 5.0.

May: Microsoft launches Windows 3.0 at $149.

May: Tim Berners-Lee re-submits proposal for WorldWideWeb.

May: AT&T assign the UNIX trademark to Unix System Laboratories (USL).

Jun: Computer Misuse Act passed in the UK.

Sep: Jean-Louis Gassée leaves Apple.

Nov: Microsoft Office 1.0 for Windows includes Word 1.1, Excel 2.0 and PowerPoint 2.0.

Dec: First web page sent across the Internet.

1991

Jan: OS/2 1.3 released, developed entirely by IBM.

Mar: NSFNET modified Acceptable Use Policy to allow commercial use.

May: Steve Jobs arranges deal for Pixar to make the film *Toy Story* with Disney.

Aug: Tim Berners-Lee writes specification for WorldWideWeb.

Sep: Mark Weiser writes article for *Scientific American* on 'ubiquitous computing'.

Oct: OS/2 2.0 announced with Workplace Shell.

Oct: Apple PowerBook released.

Oct: Apple, IBM and Motorola form the AIM alliance.

Oct: Linus Torvalds releases kernel for Linux operating system.

1992

Jan: John Sculley refers to a 'personal digital assistant' in keynote at Winter CES.

Apr: Microsoft launches Windows 3.1 with TrueType font technology.

Jun: Demon launched as first commercial Internet service provider in the UK.

Oct: Microsoft Windows for Workgroups 3.1 released.

Nov: Apple's Star Trek project demonstrates Mac operating system for PC compatibles.

1993

Jan: IBM announces losses of nearly $5 billion for the previous year.

Jan: Unix version of Mosaic web browser released by students at NCSA.

Jan: PowerPC 601 processor becomes available from AIM alliance.

Feb: Apple Macintosh Color Classic launched at $1,400.

Mar: Intel Pentium processor launched running at 50, 60 and 66MHz.

Jun: Michael Spindler replaces Sculley as Apple CEO.

Jul: Intel introduces PCI expansion bus technology.

Jul: IBM announces losses of a further $8 billion.

Jul: Microsoft launches Windows NT 3.1 at $495.

Aug: Windows for Workgroups 3.11 released with 32-bit file access.

Aug: Windows and Macintosh versions of Mosaic web browser released.

Aug: Apple Newton MessagePad launched at MacWorld Boston with starting price of $700.

Aug: The Department of Justice (DoJ) opens proceedings against Microsoft.

1994

Mar: Apple launches the Power Macintosh based on PowerPC processor and starting at $1,800.

Mar: Linux 1.0 distribution released.

Jul: Microsoft comes to an agreement with the DoJ.

Sep: Apple sues Microsoft over copyright infringements.

Sep: Cyberia, the first Internet café in the UK, opens in London.

Oct: IBM launches OS/2 3.0, also called OS/2 Warp.

Oct: World Wide Web Consortium (W3C) established with Berners-Lee as director.

Oct: Red Hat Commercial Linux released for $39.95.

Nov: Netscape Navigator web browser released.

1995

Mar: Microsoft announces 'Blackbird' and Microsoft Network (MSN.

Apr: NSFNET officially dissolved, officially allowing public use of Internet.

May: Gates sends 'The Internet Tidal Wave' memo to Microsoft staff.

Jul: Amazon website goes live.

Jul: Sun Microsystems announces the Java programming language at SunWorld 95.

Aug: Internet Explorer 1 launched as part of Microsoft Plus! For Windows 95.

Aug: Microsoft launches Windows 95 and Microsoft Network (MSN)).

Sep: The eBay website goes live.

Nov: Following premier of *Toy Story*, Steve Jobs takes Pixar public.

Dec: Microsoft holds Internet Strategy Day.

1996

Jan: Apple announces loss of $68 million over previous quarter and further job cuts.

Feb: Gil Amelio replaces Michael Spindler as CEO of Apple.

May: SuSE Linux 4.2 released.

Jun: Linux 2.0 operating system released.

Aug: Microsoft releases Internet Explorer 3.

Dec: Apple buys NeXT Software for $400 million.

1997

Jun: Netscape Communicator 4 .0 becomes available for $79.95.

Jul: Gil Amelio leaves Apple. Steve Jobs returns as interim CEO.

Aug: Microsoft purchases $150m Apple stock. Apple agrees to make IE its default browser.

Oct: DoJ asks court to hold Microsoft in contempt of the 1994 agreement.

1998

Jan: Netscape makes browser and source code available free of charge.

Feb: The Open Source Initiative (OSI) established with Eric Raymond as president.

May: DoJ files anti-trust suit against Microsoft.

May: Apple introduces the iMac, designed by Jonathan Ive, priced from $1,299.

Jun: Court of Appeals overturns initial ruling against Microsoft on browser integration.

Jun: Windows 98 launched with integrated Internet Explorer.

Sep: Google launched.

Oct: Eric Raymond receives a copy of Microsoft 'Halloween' memos.

Oct: DoJ anti-trust suit against Microsoft goes to court, Judge Jackson presiding.

1999

Feb: Windows Refund Day.

Aug: Red Hat goes public with a valuation in the region of $3 billion.

2000

Jan: Steve Jobs officially becomes CEO of Apple.

Feb: Microsoft releases Windows 2000.

Apr: Microsoft introduces Pocket PC 2000.

Jun: Judge Jackson approves DoJ's proposal to split Microsoft into two companies.

Sep: Microsoft release Windows Me (Millennium edition).

Oct: Sun releases OpenOffice under open source licence.

Nov: First crew members arrive at the International Space Station.

2001

Jan: Apple launches iTunes at MacWorld, San Francisco.

Mar: Apple launches Mac OS X, based on BSD Unix and NeXTSTEP.

Jun: Appeal court overrules Jackson and returns case to District Court.

Aug: EU expands investigation to include integration of streaming media technology.

Sep: The 9/11 terrorist attacks on America.

Oct: Microsoft releases Windows XP.

Oct: Apple launches iPod music player.

2002

Nov: Judge Colleen Kollar-Kotelly publishes final judgment of Microsoft anti-trust case.

2003

Apr: Apple launches iTunes Store selling individual songs for 99 cents.

May: Microsoft pays $750 million compensation to AOL, which now owns Netscape.

Jun: Microsoft releases Windows Mobile 2003.

Aug: Report shows Microsoft has 94 per cent of the Internet browser market.

2004

Jan: Novell acquires SuSE Linux for $210 million.

Mar: EC fines Microsoft €497m for breaching EU competition law.

Apr: Microsoft pays $1.95 billion to Sun Microsystems in compensation.

Dec: IBM announces sale of PC manufacturing business to Lenovo.

2005

Jan: Microsoft announces versions of Windows XP without Windows Media Player.

2006

Jan: Apple announces iMac and MacBook Pro machines based on Intel processors.

Apr: Microsoft appeals EC ruling.

Nov: Novell signs patent agreement with Microsoft with regards to SuSE Linux.

2007

Jan: Microsoft releases Windows Vista.

Jun: Apple iPhone released at $499.

Sep: Microsoft loses EC appeal.

Oct: OSI formally approves Microsoft's Ms-RL and Ms-PL licences as open source.

2008

Feb: EU fines Microsoft €899m for failing to comply with 2004 anti-trust order.

2009

Jul: Microsoft contributes code to the Linux kernel under the GNU GPL.

Oct: Microsoft releases Windows 7.0.

How computers work

I have tried to explain technical matters within the main text as they arise, and to avoid being unnecessarily technical. However a certain level of technical understanding is inevitably necessary when discussing a subject like this. I have therefore included this section for readers who may not be familiar with all of the terms used in the main text, or who would like to know a little more about what goes on 'under the bonnet'. Technical terms are highlighted in bold.

Representing information

A computer is a device for storing, manipulating, displaying and communicating information. This information or **data** can take many forms, but the most common are numerical values, alphanumeric text and images.

Computers are also digital devices, which means that they do their job using electronic circuits that operate much like switches, in that they can only be in one of two states: either off or on. These two states are taken to represent the binary numbers 0 and 1, which in turn are used to represent the information itself.

We are more at home with decimal numbers, based on the ten numeric symbols 0, 1, 2, 3 and so on up to 9. Once we've reached 9, we use the concept of 'place' to represent bigger values, as in 10, 11, 12. The first place, to the right, expresses individual units. The second place represents bundles of ten units at a time; the third place, hundreds of units, and so on. In this way, these ten symbols can represent just about any number that we can think of.

Binary arithmetic also uses the concept of 'place' but with only two symbols. The binary numbers 0, 1, 10 and 11 are equivalent to the decimal numbers 0, 1, 2 and 3. This means that two switches can be used to represent any decimal value between 0 and 3. Add a third switch, or **bit**, and we can count from 000 to 111 (0 to 7 in decimal). Add a fourth and we can count to 1111, or 15 in decimal. Such a device is known as a 4-bit device and is described as having a **word length** of 4 bits. The earliest microprocessors were 4-bit devices; however these were soon replaced by 8-bit devices, capable of handling decimal values from 0 to 255 in one go. At the time of writing, microprocessors that have a word length of 64 bits are replacing the

32-bit devices that appeared in 1984. A 64-bit device is capable of handling decimal values of nearly 19 billion billion in one go.

Binary numbers represent alphanumeric characters through the use of a code. A very important code in the computer world is the American Standard Code for Information Interchange (ASCII), which was published in 1963. ASCII uses 7-bit values to define 128 characters including both upper and lower case letters, the digits 0 to 9, a selection of punctuation marks and a number of 'control' characters. A lower case 'a', for example, is represented by the binary number 1100001, the numeric character '3' by 0110011 and the dollar sign by 0100100.

Although ASCII is a 7-bit code, in the early days it was usual to add an eighth bit, called the 'parity' bit, to help to detect errors in communication. Because computers spent so much time dealing with text, the term **byte** became synonymous with an 8-bit value. It inevitably followed that a 4-bit value became known as a 'nibble', or (for the sake of consistency) a 'nybble'.

ASCII has now largely been replaced by Unicode, which was introduced in 1991 and uses 16-bit or even 32-bit values to represent more than 100,000 different characters, including those found in Arabic, Chinese, Japanese and many other alphabets, as well as a wide range of symbols.

Images are represented in a computer as a matrix of dots or **pixels**, each one of which can be displayed in a range of colours. The range of colours is determined by the number of bits used to express each pixel. The original Apple Macintosh, for example, had a screen resolution of 512 by 384 pixels but used just one bit for each pixel, which could therefore be either black or white. The screen on a modern desktop computer can typically display 1,920 by 1,080 pixels (the same resolution as a wide screen High Definition television), with each pixel allocated a 24-bit value. This is enough to express over 16 million colours.

Of course to the computer all of this data is simply binary code. It is the computer program that determines whether one binary value represents an alphanumeric character or a pixel that is part of an image.

All about memory

The two most important parts of a computer are the central processing unit (CPU), which is responsible for actually processing data and moving it about, and the memory, which stores the data and the program of instructions telling the processor what to do with it. The memory of a computer works much like a pigeon hole system of the sort you might find in the lobby of a hotel. Each pigeon hole in the store holds one program instruction or data value, and has a capacity of typically 4, 8, 16, 32 or 64 bits, depending on the word length of the device. The pigeon holes themselves are identified by their addresses, which are also expressed as binary numbers.

The capacity of a memory store is measured in terms of the number of bytes of data that it can uniquely address, so in theory a memory capable of storing 1,000 bytes of information should be referred to as having a capacity of one kilobyte. However, as these are binary devices, it is more common to refer to a capacity of two to the power of ten (2^{10}) or 1,024 bytes as one **kilobyte** (usually shortened to 1KB). A capacity of 1024^2 or 1,048,576 is referred to as 1 **megabyte** or 1MB; while 1024^3, which comes to a little over a billion, is 1 **gigabyte** or 1GB.

The amount of memory that a processor can address is determined by the number of wires that connect the processor to the memory; or the 'width of the address bus', to put it in a more technical fashion. A 32-bit address bus, for example, can access up to 2^{32} or 4GB of memory. However earlier microprocessors with narrower address buses had to play tricks in order to access usable amounts of memory.

The Intel 4004, for example, was a 4-bit microprocessor, so it connected to the outside world through a 4-bit address bus, which in theory would restrict it to just 16 separate memory addresses. However a technique known as 'multiplexing' allowed it to use 12-bit addresses, giving it access to a more useful 4,096 bytes or 4KB of memory. This was achieved by splitting each address into three separate 4-bit values and sending them sequentially to the memory chips, where another chip would reconstitute the original 12-bit address. Intel did this because it was anxious to keep the number of connectors on the chip down, but the trade-off was that it took three times as long to transfer each address.

The Intel 8086, the company's first fully 16-bit microprocessor and the basis of many PC compatibles, actually had a 20-bit external address bus which gave it access to a maximum of 1MB of memory. This was achieved by combining a 16-bit 'offset address' with a 16-bit 'segment address'. First the segment address was turned into a 20-bit number by shifting it four bits to the left, which has the effect of multiplying it by 2^4 or 16, and then the result was added to the offset address. Such a technique might seem cumbersome, but did have its advantages. Each segment address effectively referenced a 64KB block within the memory, conveniently allowing a program to allocate particular segments to particular parts of the program or particular types of data.

Memory chips can implement Read Only Memory **(ROM)**, where the data stored in the chip cannot easily be changed; or Random Access Memory **(RAM)**, where data can be read or written with ease. ROM chips are generally used to store the program code that defines the function of a device, and so does not need to be changed once it has been written. ROM chips are generally described as 'non-volatile', because their content is retained even when the chip is disconnected from a power supply. RAM chips tend to be much faster, but are 'volatile', as content is only retained while the chip is powered.

RAM chips can in turn be either 'static' (known as **SRAM**) or 'dynamic' **(DRAM)**. The earliest memory chips were static. Although reliable, fast and easy to use, SRAM requires at least six transistors for each memory cell which takes up a lot of space on the chip surface and requires a lot of power. By contrast, DRAM stores each bit of data as an electrical charge across the plates of a capacitor. A single memory cell can be built using just one transistor and one capacitor, which makes DRAM much cheaper, less power-hungry and capable of storing much more data. However the charge across the capacitor leaks, which means that it has to be refreshed thousands of times a second (hence the term 'dynamic'). Refreshing the charge requires external support circuitry, which makes DRAM harder to use. These days, the internal memory of virtually all computers is DRAM.

The ideal computer memory is cheap and fast and retains its content forever. Unfortunately, achieving all three with a single technology has so far proved impossible. Conventional memory chips are very fast, but are relatively expensive and retain data only as long as they remain connected to a power source. Flash memory (as used in today's USB memory sticks) will retain data without a power source, but is slower and even more expensive. Disk drives are much cheaper in terms of cost per megabyte, but also much slower. As a result, computers rely on a hierarchy of memory, moving data from fast but volatile memory chips onto slower, less volatile disk drives as the need arises.

Inside the microprocessor

A microprocessor is the implementation of the Central Processing Unit (CPU) of a computer as a single integrated circuit. As such it is responsible for manipulating data in accordance with a sequence of instructions, known as the program, and returning the results. The instructions themselves are also expressed as binary numbers, and are stored alongside the data in the computer's memory. In addition to the memory, the CPU is connected to circuitry that handles the devices that it needs to communicate with the outside world, such as a keyboard, screen or printer, and to a clock that determines the rate at which instructions are processed.

The internal architecture of a modern microprocessor is extremely complicated. However the essential elements include the Arithmetic and Logic Unit (ALU), where operations are actually carried out, and the various registers that act as high-speed 'scratch pad' memories. Each register can store a single binary word, and the fastest is the Accumulator, which is directly connected to one of the ALU inputs.

Also important is the Program Counter, which contains the address of the next instruction to be processed, and the Instruction Register, which contains the actual bits that make up the instruction that is currently being processed. All of these components are connected together by an internal data bus which, together with the address bus, connect to the outside world through pins on the microprocessor housing. Other elements include the Instruction Decoder, which understands how to interpret the instructions, and the Timing and Control Unit, which orchestrates their execution.

The best way to understand the operation of a microprocessor is to follow what happens when a single instruction is fetched from external memory, decoded and executed:

1. The program counter contains the location in the memory of the instruction to be executed, so its contents are transferred to the external address bus. It takes a little time for the result to return, so in the meantime, the microprocessor increments the program counter by one so that it is ready with the address of the next instruction.

2. The content of the requested location returns through the data bus and is directed into the instruction register by the control unit.

3. For the purposes of illustration, we will assume the instruction directs the microprocessor to add a value located in memory to a value already stored in the accumulator by an earlier instruction. The memory address of the required value has likewise been stored in another register, so the content of this register is transferred to the address bus.

4. The content of the requested location returns and is directed to the ALU, where it is added to the contents of the accumulator. The result is put back

into the accumulator, where it remains, ready to be acted upon by the next instruction.

It is important to note that fetching, decoding and executing this single instruction involves a considerable number of operations, taking up a number of clock cycles. This is why it is dangerous to measure the speed of a microprocessor solely by its clock rate: the efficiency with which instructions are carried out is just as important. The 8086 processor, for example, takes around 12 clock cycles to execute a single instruction, while the 80286 and 80386 both take less than five.

These days, microprocessors employ a wide range of tricks to speed things up. One common technique is **pipelining,** which makes use of the fact that, for most of these operations, only part of the microprocessor is actually being used while the rest lays idle. In a system that employs a pipeline, the next instruction is being loaded from memory at the same time as the current instruction is executing and the previous one is writing its result into the appropriate register. Pipelining requires careful coordination by the control unit, but can be very effective. Such techniques allow more advanced processors such as the Intel Pentium II and III to average three or four instructions per clock cycle (Mueller & Soper, 2001).

All modern microprocessors also carry considerable amounts of memory on the chip itself. This is not to replace external memory, but rather to act as a **cache**, mirroring external memory to speed up storage and retrieval operations. Indeed modern microprocessors have several levels of cache: an Intel Pentium 4, for example, could have up to 16KB of Level 1 cache and 2MB of Level 2 cache, while still being connected to gigabytes of external memory.

Computer programs

The microprocessors that are at the heart of our computers understand binary code. Each microprocessor comes with a predefined instruction set – a set of binary codes that it will interpret in predefined ways. The Intel 8080, for example, was manufactured to understand over 80 different instruction codes, each assigned a unique 8-bit value. The instruction 10000011 tells the processor to add the contents of its E register to the contents of its accumulator, and write the result back into the accumulator. The instruction 10000001 tells it to do the same, but using the C register instead of the E register.

A computer program consists of thousands or even millions of such instructions stored sequentially within the computer's memory. Such programs are said to be written in **machine code**, because that is the language that the machine understands.

However machine code is notoriously difficult to read or write, even for geeks. More comprehensible is a program expressed in **assembly language**, where the binary codes are replaced with more intuitive mnemonics. Written in assembly language, the instruction 10000011 becomes 'ADD E' while 10000001 becomes 'ADD C'. In this way, the process of adding two numbers together can be written as:

```
MVI A, 3
MVI B, 4
ADD B
```

What this program does is write the value '3' into the microprocessor's accumulator (referred to as 'A'), write the value '4' into the B register, add the contents of the B register to the contents of the accumulator and update the accumulator with the result. Running such a program involves entering these instructions into another program, called an **assembler**, which replaces the mnemonics with the binary machine code that the computer understands.

Machine code and assembly language are described as low-level languages, because they directly describe what the CPU must do in order to process the instruction. If you want your program to run as fast as possible and take up as little space as possible, then you need to write it in assembly language. However as microprocessors got faster and memory became cheaper, such considerations became less important, while other factors, such as readability and portability (the ability to run the same program on computers made by different manufacturers), became more significant. This is where high-level languages such as COBOL, FORTRAN, C and BASIC come in, offering a higher level of abstraction.

This is what our assembly language program might look like, written in BASIC:

```
let a = 3
let b = 4
let a = a + b
```

At first glance the BASIC version looks much the same, if a little easier to understand, as the assembly language. However there is a big difference, because the letters 'a' and 'b' do not refer to specific registers within the CPU, but rather to abstract variables. Behind the scenes, another program (either a **compiler** or an **interpreter**) makes decisions about how best to execute the program and translates it into machine code instructions that the microprocessor can understand, but this is not the concern of the high-level programmer. All the programmer needs to know is that he now has a variable labelled 'a' which he can reference later in the program, confident that it will have a value of '7'. He doesn't care where in the computer it is actually stored.

High level languages really come into their own in real-world situations. The following BASIC program, for example, prints out a table of Fahrenheit temperatures and their Celsius equivalents:

```
for f = 0 to 300 step 20
  c = (5/9) * (f - 32)
  print f, c
next f
```

Again, its operation is fairly self-explanatory, even to the layperson. However its equivalent in assembly language would require considerably more lines and would be incomprehensible to all but the most expert. In particular, the 'print' instruction would require the numeric values represented by 'f' and 'c' to be translated into alphanumeric characters and transferred to the circuitry responsible for displaying them on screen or printing them out on a printer.

Although they work at a higher level of abstraction, languages such as BASIC or C still tell the computer what procedures to use to solve the problem. For this reason, they are described as procedural or imperative languages. They are also referred to as 'third generation' languages or 3GLs, on the basis that machine code is first generation, assembly language is second generation and procedural languages are third generation.

There is some debate as to what constitutes a fourth generation language (4GL), although logically such a language should work at an even higher level of abstraction. Some claim that a 4GL should state the problem, but leave it to the computer to work out how best to solve it. A well-known example would be Structured Query Language (SQL), which is used to extract information from databases. For example:

```
SELECT name, salary FROM employee WHERE age > 50
```

This SQL statement tells the computer to select the names and salaries of everyone listed as an employee who is more than 50 years of age. It is quite specific about what it wants, but leaves it to the computer to work out how best to execute the instruction.

Others have made claims for various domain-specific languages to be described as 4GL. As the name suggests, a domain-specific language (DSL) applies a higher level of abstraction to a particular class or 'domain' of problems. SQL is a DSL, in that it is solely concerned with the extraction of data from databases.

Acknowledgements

In addition to the sources listed in the Bibliography, I should also mention the Wikipedia website, which proved invaluable in guiding me to many of the original documents, interview transcripts and other material that I have referenced. Also invaluable were the Internet Archive, which includes not only documents but also video and audio recordings; the Charles Babbage Institute Collections which contain over 300 oral history transcripts; the 1000Bit website with its thousands of original brochures, magazine articles and manuals; The Online Books Page with links to the full text of over a million books and magazines; and the DigiBarn computer museum with its collection of manuals, videos and interviews.

My thanks go to my editors Paul Stephens and Liz Broomfield; to Colleen Lanchester-Raynie for a great cover (and to 99designs for making it possible); to Joanna Penn and Danuta Kean for introducing me to both sides of the book publishing business; to all the journalists, writers and editors that I have met, worked and shared a drink with in the course of my career; and to Hazel, Jemma and Luke who make it all worthwhile.

Note that all estimates of the current value of a past price have been made using the calculators at the MeasuringWorth website.

About the author

MATT NICHOLSON was born in south east England in the early 1950s. His first encounter with a computer was as a member of a sixth-form technology club. The club had an arrangement with a local company such that programs written by the students on a teleprinter terminal would be run on the company's mainframe, and the results (usually a list of errors) returned a few days later. This was followed at university in the early 1970s where, as part of his Computer Science course, he was given the task of writing a component of a magnetic disk storage system in assembly language for an already outdated LEO III that the university had bought for educational purposes.

Matt went on to become a technical journalist at Haymarket Publishing, where he became editor of the country's top-selling *What Hi-fi* magazine. Bored with hi-fi, he left Haymarket to join VNU Business Publications in 1983 as editor of *What Micro*, the county's first 'buyers guide' magazine for the microcomputer industry. In 1986 he left VNU to become Future Publishing's ninth employee, editing the company's first magazine, *Amstrad Action*. Later the same year Matt became launch editor of *PC Plus* which initially concentrated on the low-cost Amstrad PC but went on to become Britain's biggest general-interest PC magazine.

Leaving Future Publishing in 1989, Matt pursued a freelance career, writing articles for a wide range of computer magazines as well as for the computer sections of the *Guardian* newspaper and the *Financial Times*. He also worked for EMAP as a consultant editor of *What Personal Computer*. In 1997 he set up Matt Publishing to publish *Developer Network Journal* for Microsoft, and in 2003 became editor of *HardCopy*, the customer magazine of UK software distributor Grey Matter. Matt lives in Bristol in the south west of England.

Bibliography

Adamson, I., & Kennedy, R. (1986). *Sinclair and the 'Sunrise' Technology.* Penguin Books.

alistairw. (2006, July 19). *GameSetInterview: Adventure International's Scott Adams.* Retrieved June 23, 2011, from Game Set Watch: http://www.gamesetwatch.com/2006/07/gamesetinterview_adventure_int.php

Alsop, S. (1993, August 30). Newton deserved a better introduction, not a warmer reception. *InfoWorld,* p. 4.

Anchordoguy, M. (1989). *Computers Inc.: Japan's challenge to IBM.* Cambridge, Massachusetts: Harvard University Press.

Andersen, P. (Director). (1984, December 13). *Commercial Breaks (BBC2)* [Motion Picture].

Anderson, C. (2012, April 24). *The Man Who Makes the Future.* Retrieved October 11, 2012, from Wired: http://www.wired.com/business/2012/04/ff_andreessen/

Anderson, J. J. (1984, July). Apple Machintosh; cutting through the ballyhoo. *Creative Computing,* p. 12.

Anderson, T., & Galley, S. (1985). The History of Zork. *The New Zork times,* 4(1-3).

Angwin, J. (1998, January 14). U.S. Judge Scolds Microsoft / Firm may wind up in contempt over tactics. *San Francisco Chronicle.*

Antonoff, M. (1991, February). Gilbert Who? *Popular Science,* pp. 70-73.

Apple. (2012, February 19). *Macintosh Portable: Used in Space Shuttle.* Retrieved from Apple: http://support.apple.com/kb/TA30635

Apple Computer. (1990, May 28). Now that everyone agrees how a computer should work, try one that actually works that way. *InfoWorld,* pp. S24-25.

Apple Computer. (1999, January 5). 800,000 iMacs Sold in First 139 Days. San Francisco.

Apple vs. Microsoft, 35 F.3d 1435 (9th Cir. 1994) (1994).

Arthur, C. (2012). *Digital Wars: Apple, Google, Microsoft & the battle for the Internet.* London: Kogan Page.

Atack, C. (1988, October). From Atom to Arc. *Acorn User.*

Atkinson, B. (2010, June 4). *The Genius of Design 5: Objects of Desire.* BBC2.

Ballmer, S. (1996, June). Triumph of the Nerds: Part II. (R. X. Cringely, Interviewer) PBS.

Banks, W. C. (1987, July 1). Desktop Publishing: The advent of laser-driven printers and page-layout programs has opened the publishing world to the era of the entrepreneur. *Money Magazine.*

Bartholomew, D. (2004, December 21). *What's Really Driving Apple's Recovery.* Retrieved May 14, 2013, from IndustryWeek:
http://www.industryweek.com/articles/whats_really_driving_apples_recovery_325.aspx

Bartimo, J., & McCarthy, M. (1985, February 11). Is Apple's Laserwriter On Target? *InfoWorld,* pp. 15-18.

Bateman, S. (1985, May). GEM: A New Look For IBM And Atari. *Compute!*(60), p. 22.

Baum, D. (1994, May 9). Power Mac Migration. *InfoWorld,* p. 70.

Berners-Lee, T. (1999). *Weaving the Web: The Past, Present and Future of the World Wide Web by its Inventor.* London: Orion Business Books.

Berners-Lee, T. (2001, August 8). Electric Journeys. (S. Fogarty, Interviewer)

Berners-Lee, T. (n.d.). *Frequently Asked Questions.* Retrieved October 10, 2012, from World Wide Web Consortium (W3C): http://www.w3.org/People/Berners-Lee/FAQ.html

Blundon, W. (1996, October 1). Netscape outlines its vision for the future: NONE. *JavaWorld.*

Boone, G. W. (1973). *Patent No. 3,757,306.* United States of America.

Both, D. P. (1997). *A Short History of OS/2.* Retrieved January 13, 2012, from Norloff's OS/2BBS.COM: http://www.os2bbs.com/os2news/os2history.html

Box Office Mojo. (n.d.). *1995 Domestic Grosses.* Retrieved April 5, 2013, from Box Office Mojo: http://boxofficemojo.com/yearly/chart/?yr=1995

Branwyn, G. (2010, January 9). *Alt.CES: Andy Warhol at the Amiga launch, 1985.* Retrieved June 5, 2013, from Make:: http://blog.makezine.com/2010/01/09/altces-andy-warhol-at-the-amiga-lau/

Brick, M. (2001, June 28). U.S. Appeals Court Overturns Microsoft Breakup Ruling. *The New York Times.*

Brinkley, J., & Lohr, S. (2000, June 9). *Retracing the Missteps in Microsoft's Defense at Its Antitrust Trial.* Retrieved November 21, 2012, from Law Offices of C. Richard Noble: http://www.richardnoble.com/microsoft-trial.htm

Brodie, R. (2000, May 3). *Microsoft: The Early Days.* Retrieved May 24, 2012, from Meme Central: http://www.memecentral.com/mylife.htm

Brown, J., Goldman, M., Gunter, T., Shahan, V., & Walker, B. (2007, July 23). Oral History Panel on the Development and Promotion of the Motorola 68000. (D. House, Interviewer) Computer History Museum.

Brown, S. (2008, July 21). WarGames: A Look Back at the Film That Turned Geeks and Phreaks Into Stars. *WIRED*.

Bunnell, D. (1975, August). Across the Editor's Desk. *Computer Notes, 1*(3).

Burkeman, O. (2009, October 23). The Day The World Changed. *The Guardian: The Internet Turns 40*.

Caldera Inc. (1996, July 24). *Software Developer Caldera Sues Microsoft For Antitrust Practices Alleges Monopolistic Acts Shut Its DR DOS Operating System Out of Market*. Retrieved August 16, 2012, from MaxFrame: http://www.maxframe.com/DR/Info/fullstory/ca_sues_ms.html

Call MD Plus. (2012, May 2). *Rowland Hanson: The Windows Story*. Retrieved May 29, 2012, from http://www.youtube.com/watch?v=6IlaNPOh1e0

Carlton, J. (1997). *Apple: The Inside Story of Intrigue, Egomania, and Business Blunders*. New York: Random House.

Cartwright, W., Peterson, M. P., & Gartner, G. (Eds.). (1999). *Multimedia Cartography*. Berlin Heidelberg: Springer-Verlag.

Chandrasekaran, R., & Leibovich, M. (1999, January 15). Microsoft Witness Recants Monopoly View. *Washington Post*, p. E3.

Chposky, J., & Leonsis, T. (1988). *Blue Magic: The People, The Power and the Politics behind the IBM Personal Computer*. London: Grafton Books.

Christensen, W., & Suess, R. (1989). *The Birth of the BBS*. Retrieved August 29, 2012, from Chinet -- Public Access since 1982: http://chinet.com/html/cbbs.html

Clough, B., & Mungo, P. (1992). *Approaching Zero: Data Crime and the Computer Underworld*. London: faber and faber.

Commodore History. (1989). *You Don't Know Jack!* Retrieved July 13, 2012, from http://www.commodore.ca/history/people/1989_you_dont_know_jack.htm

CompuServe. (1984, April). Electronic Shopping Mall. *Online Today*.

Computer History Museum. (1997). *Exhibits: A History of the Internet 1962-1992*. Retrieved September 17, 2012, from Computer History Museum: http://www.computerhistory.org/internet_history/internet_history_80s.html

Condon, C. (1998, December 7). *Statement by Attorney General Condon on Microsoft*. Retrieved November 27, 2012, from Tech Law Journal: http://www.techlawjournal.com/courts/dojvmsft2/81207.htm

Connick, J. (1986, October). Steve Wozniak. *Call-A.P.P.L.E.*

Cook, K. (1984, June 26). Digital Research Ties CP/M, DOS. *PC Magazine*, p. 39.

Cornell University Law School. (1992, October 23). *42 USC § 1862 - Functions*. Retrieved September 21, 2012, from Legal Information Institute: http://www.law.cornell.edu/uscode/text/42/1862

Crabb, D. (1995, December). Copland, Where Are Thee? *MacTech, 11*(12).

Cringely, R. X. (1992). *Accidental Empires*. Viking.

Cringely, R. X. (Director). (1998). *Nerds 2.0.1: Wiring the World* [Motion Picture].

Current, M. D. (2011, January 13). *A History of the Former Atari Japan*. Retrieved February 22, 2011, from Michael Current: http://mcurrent.name/atarihistory/japan.html

Dale, R. (1986, January). The Sinclair Story Part 2. *Sinclair User*(46).

Dalziel, W., Sollman, G., & Massaro, D. (2005, January 3). Oral History Panel on 5.25 and 3.5 inch Floppy Drives. (J. Porter, Interviewer) Mountain View, California: Computer History Museum.

Danger Mouse. (1995, July 26). *Allabout Callahan's*. Retrieved September 13, 2012, from faqs.org: http://www.faqs.org/faqs/callahans/allabout/part1/

DeLamartar, R. T. (1986). *Big Blue: IBM's use and abuse of power*. New York: Dodd, Mead & Company, Inc.

Department of Justice. (1998, December 15). *Antitrust Case Filings*. Retrieved November 22, 2012, from The United States Department of Justice: http://www.justice.gov/atr/cases/f2000/gates6.pdf

Department of Justice. (1998, May 18). *Justice Department Files Antitrust Suit Against Microsoft for Unlawfully Monopolizing Computer Software Markets*. Retrieved November 22, 2012, from The United States Department of Justice: http://www.justice.gov/opa/pr/1998/May/223.htm.html

Digital Research. (1984, May). Concurrent DOS bridges PC-DOS, CP/M. *Digital Research News, 4*(2), p. 3.

Discovery Communications. (2012). *John Sculley: Apple's Early Years*. Retrieved March 8, 2013, from http://dsc.discovery.com/tv-shows/curiosity/topics/john-sculley-apples-early-years.htm

Dobrzynski, J. H. (1993, February 1). IBM's Board Should Clean Out The Corner Office. *BusinessWeek*.

Dompier, S. (1975, May). Music of a Sort. *People's Computer Company*, p. 8.

Dvorak, J. C. (1996, October 22). Inside Track. *PC Magazine*.

Edge Staff. (2009, May 22). *The Making Of: Elite*. Retrieved July 29, 2011, from EDGE: http://www.next-gen.biz/features/making-elite?page=2

Edwards, B. (2008, August 15). Eight Ways the IMac Changed Computing. *MacWorld*.

Edwards, B. (2011, January 17). *Happy 25th Birthday, Mac Plus*. Retrieved January 8, 2013, from Macworld Australia: http://www.macworld.com.au/news/happy-25th-birthday-mac-plus-22027/

Faggin, F. (1992, March). The Birth of the Microprocessor. *Byte*, p. 145.

Faggin, F., Feeney, H., Gelbach, E., Hoff, T., Mazor, S., & Smith, H. (2006, September 21). Oral History on the Development and Promotion of the Intel 8008 Microprocessor. (D. House, Interviewer) Mountain View, California, USA: Computer History Museum.

Faggin, F., Shima, M., & Ungermann, R. (2007, April 27). Oral History Panel on the Founding of the Company and the Development of the Z80 Microprocessor. (M. Slater, Interviewer) Mountain View, California, USA: Computer History Museum.

Fairhead, H. (1992). *The 386/486 PC: Second Edition*. Leyburn, North Yorkshire: I/O Press.

Felsenstein, L. (1996, June). Triumph of the Nerds. (R. X. Cringely, Interviewer)

Ferguson, C. H., & Morris, C. R. (1994). *Computer Wars: The Fall of IBM and the Future of Global Technology*. New York: Times Books.

Fischer, E. (n.d.). *The Evolution of Character Codes, 1874-1968*. Retrieved June 11, 2010, from http://www.transbay.net/~enf/ascii/ascii.pdf

Fontana, J. (2009, July 20). *Microsoft stuns Linux world, submits source code for kernel*. Retrieved August 1, 2013, from Network World: http://www.networkworld.com/news/2009/072009-microsoft-linux-source-code.html

Fothergill, R. (1981, April). *The Microelectronics Education Programme Strategy*. Retrieved from National Archive of Educational Computing: http://www.naec.org.uk/organisations/the-microelectronics-education-programme/the-microelectronics-education-programme-strategy

Free Software Foundation. (2007, June 29). *GNU General Public License*. Retrieved from GNU Operating System: http://www.gnu.org/licenses/gpl.html

Free Software Foundation. (2013, April 10). *Free GNU/Linux distributions*. Retrieved July 13, 2013, from GNU Operating System: http://www.gnu.org/distros/free-distros.html

Galletti, C. (2007, April). *Origins of S-100 computers*. Retrieved October 22, 2010, from retrotechnology.com: http://retrotechnology.com/herbs_stuff/s_origins.html#s100

Gan, D. (2011, July 22). *The Internet Tidal Wave*. Retrieved October 4, 2012, from Letters of Note: http://www.lettersofnote.com/2011/07/internet-tidal-wave.html

Gantz, J. (1985, March 11). The Revolution In In-House Publishing. *InfoWorld*, p. 32.

Gassée, J.-L. (2010, January 17). *The Apple Licensing Myth.* Retrieved January 11, 2013, from Monday Note: http://www.mondaynote.com/2010/01/17/the-apple-licensing-myth/

Gassée, J.-L. (2012, October 15). Gut-level Affection, An interview with Jean-Louis Gassée. (H. Dediu, Interviewer)

Gates, B. (1976, April 10). Open Letter to Hobbyists. *dr dobb's journal of Computer Calisthenics & Orthodontia.*

Gates, B. (1983). Steve Jobs hosts the Macintosh dating game at the Macintosh pre-launch event. (S. Jobs, Interviewer)

Gates, B. (1993). Bill Gates Interview. (D. Allison, Interviewer) National Museum of American History, Smithsonian Institution.

Gates, D. (2000, May 11). *Microsoft's Kerberos shuck and jive.* Retrieved July 25, 2013, from The Industry Standard: http://www.networkworld.com/news/2000/0511kerberos.html

Goldberg, A. (1996, June). Triumph of the Nerds: Part III. (R. X. Cringely, Interviewer) PBS.

Goldstine, H. H. (1992). Computers at the University of Pennsylvania's Moore School, 1943-1946. *Proceedings of the American Philosophical Society, 136*(1).

Goodell, J. (1994, June 16). Steve Jobs: The Rolling Stones Interview. *Rolling Stone, 684.*

Goodin, D. (1999, April 28). *Microsoft emails focus on DR-DOS threat.* Retrieved August 16, 2012, from CNET News: http://news.cnet.com/2100-1001-225129.html

Gore, B. (1982, March). *Your Computer, 2*(3), p. 24.

Government Information Office, Republic of China (Taiwan). (1984, May 27). IBM Accepts Redesign Of Computer System. *Taiwan Today.*

Graham, F. (2013, April 11). *Interview: The man behind the PowerPoint presentations.* Retrieved September 27, 2013, from BBC News: http://www.bbc.co.uk/news/business-22110435

Gray, S. B. (1984, November). The Early Days of Personal computers. *Creative Computing, 10*(11), p. 6.

Hawker, N. (2003, August 5). *A Background Briefing on the European Microsoft Case.* Retrieved December 20, 2012, from American Antitrust Institute: http://www.antitrustinstitute.org/~antitrust/node/10290

Heilemann, J. (2000, November). The Truth, The Whole Truth, And Nothing But The Truth: The Untold Story of the Microsoft antitrust Case. *Wired,* pp. 261-311.

Heise, V. H. (1998, August 7). *Linus Torvalds Talks about Current Linux Issues.* Retrieved July 18, 2013, from Anonymous Insider: http://www.anonymous-insider.net/linux/research/1998/0807.html

Hertzfeld, A. (2005). *Revolution in the Valley.* Sebastopol, CA: O'Reilly Media, Inc.

Hiltzik, M. A. (2000). *Dealers of Lightning: Xerox PARC and the Dawn of the Computer Age.* New York: HarperBusiness.

Hoffman, J. (1981, July 12). *Preliminary Macintosh Business Plan [DRAFT].* Retrieved March 21, 2012, from Computer History Museum: http://www.computerhistory.org/collections/accession/102712692

Holt, R. M. (1998, September 22). The F14A Central Air Data Computer and the LSI Technology State-of-the-Art in 1968.

Hughes, P. (1994, April 1). Interview with Patrick Volkerding. *Linux Journal.*

IBM. (1987, April 2). *IBM OS2 1.0 announcement.* Retrieved July 25, 2012, from Computer History Wiki: http://gunkies.org/wiki/IBM_OS2_1.0_announcement

IBM. (2011). *The PC: Personal Computing Comes of Age.* Retrieved November 25, 2013, from IBM at 100: http://www-03.ibm.com/ibm/history/ibm100/us/en/icons/personalcomputer/

IBM. (n.d.). *The birth of the IBM PC.* Retrieved November 28, 2011, from IBM Archives: http://www-03.ibm.com/ibm/history/exhibits/pc25/pc25_birth.html

IDC. (2013, February 14). *Android and iOS Combine for 91.1% of the Worldwide Smartphone OS Market in 4Q12 and 87.6% for the Year, According to IDC.* Retrieved October 2, 2013, from IDC: http://www.idc.com/getdoc.jsp?containerId=prUS23946013

InfoWorld staff. (1987, March 2). Industry Leaders Speculate on the Importance of New Macs. *InfoWorld, 9*(9), p. 17.

Intel, NCM. (1970, February 6). *Agreement between Intel & NCM.* Retrieved April 23, 2010, from Xnumber: http://www.xnumber.com/xnumber/agreement.htm

Internet Society. (2012). *Brief History of the Internet.* Retrieved September 21, 2012, from http://www.internetsociety.org/internet/internet-51/history-internet/brief-history-internet

Isaacson, W. (2011). *Steve Jobs: The Exclusive Biography.* London: Hachette Digital.

Jerz, D. G. (2007, Summer). *Somewhere Nearby is Colossal Cave: Examining Will Crowther's Original 'Adventure' in Code and in Kentucky.* Retrieved June 23, 2011, from digital humanities quarterly: http://www.digitalhumanities.org/dhq/vol/1/2/000009/000009.html

Jobs, S. (1996, June). Triumph of the Nerds. (R. X. Cringely, Interviewer) PBS.

Jobs, S. (1996, June). Triumph of the Nerds: The Rise of Accidental Empires. (R. X. Cringely, Interviewer) PBS.

Jobs, S. (2010, June 7). The iPad: Past, Present, Future. (W. Mossberg, & K. Swisher, Interviewers) The Wall Street Journal.

Johnson, H. R. (2009, July 1). *Jon Titus, Bugbooks, and the Mark 8.* Retrieved June 30, 2010, from retrotechnology.com: http://www.retrotechnology.com/dri/titus.html

Johnston, S. (1991, July 8). Microsoft drops OS/2 2.0 API, revamps 32-bit Windows plans: Users face choice between OS/2 and Windows NT. *InfoWorld,* p. 1.

Johnstone, B. (2003). *Never Mind the Laptops: Kids, Computers and the Transformation of Learning.* Lincoln, NE: iUniverse Inc.

Jones, D. W. (2001, June 8). *PDP-8 Frequently Asked Questions.* Retrieved March 3, 2010, from Internet FAQ Archives: http://www.faqs.org/faqs/dec-faq/pdp8/

Kampert, P. (2003, February 16). Low-key pioneer. *Chicago Tribune.*

Kane, M. (1998, July 8). *Jobs: Apple still on the right track.* Retrieved May 14, 2013, from ZDNet: http://www.zdnet.com/news/jobs-apple-still-on-right-track/99946

Karlgaard, R. (2013, September 2). Ahead Of Their Time: Noble Flops. *Forbes.*

Kass, M. (n.d.). *DigiBarn Books: BASIC Computer Games by David Ahl (1978, 1973).* Retrieved August 14, 2010, from DigiBarn Computer Museum: http://www.digibarn.com/collections/books/basicgames/#first

Kay, A. C. (1969). *The Reactive Engine.* University of Utah.

Kay, A. C. (1972). A Personal Computer for Children of All Ages. *Proceedings of the ACM National Conference.* Boston.

Kay, A. C. (1993). *The Early History of Smalltalk.* Retrieved Oct 13, 2013, from Gagne: Technology and Research Archive: http://gagne.homedns.org/~tgagne/contrib/EarlyHistoryST.html

Kay, A. C., & Goldberg, A. (1977, March). Personal Dynamic Media. *Computer.*

Kearns, D. (1997, March 31). Gang of Four goes after Microsoft. *Network World,* p. 28.

Kehoe, L. (1997, February 25). Doubts Grow about Leadership at Apple. *Financial Times.*

Kelly, D. (1995). *Handout for the UNIX Industry: A Brief History.* Retrieved May 27, 2013, from Community Resource and Instructor Support Program: http://snap.nlc.dcccd.edu/learn/drkelly/brf-hist.htm

Kilby, J. S. (1959, Feb 6). *Patent No. 3,138,743.* United States of America.

Kildall, S. (1997, July 1). *An Homage to Gary Kildall.* Retrieved September 30, 2013, from Digital Research: http://www.digitalresearch.biz/DR/Gary/9775b.htm

Knight, G. (2006). *Commodore-Amiga Sales Figures.* Retrieved July 17, 2012, from Amiga History Guide: http://www.amigau.com/aig/sales.html

Kohler, C. (2010, May 21). *Q&A: Pac-Man Creator Reflects on 30 Years of Dot-Eating.* Retrieved February 23, 2011, from WIRED: http://www.wired.com/gamelife/2010/05/pac-man-30-years/

Kouyoumdjian, V. (1994, August). *DOS/V, Windows, Prices, and the Future...* Retrieved August 15, 2012, from Computing Japan: http://www.japaninc.com/cpj/magazine/issues/1994/aug94/08trend.html

Lammers, S. (1986). *Programmers at Work.* Redmond, Washington: Microsoft Press.

Lampson, B. (1972, December 19). *Why Alto? Butler Lampson's Historic 1972 Memo.* Retrieved March 5, 2012, from DigiBarn computer museum: http://www.digibarn.com/friends/butler-lampson/

Lampson, B. (2006, October 18). The Alto and Ethernet System - Xerox PARC in the 1970s. Seattle, WA, USA: University of Washington: Computer Science and Engineering.

LaPlante, A. (1990, May 28). Users Have High Hopes for 3.0. *InfoWorld, 12*(22), p. S39.

Lasar, M. (2012, May 31). *25 years of HyperCard—the missing link to the Web.* Retrieved September 22, 2012, from Ars Technica: http://arstechnica.com/apple/2012/05/25-years-of-hypercard-the-missing-link-to-the-web/

Lazzareschi, C. (1989, February 19). APPLE: HAS IT LOST ITS BITE? *Los Angeles times.*

King, M. (Producer), & Lee, R. (Director). (2001). *The Secret History of Hacking* [Motion Picture]. Channel 4.

Lemmons, P. (1981, October). The IBM Personal Computer: First Impressions. *BYTE*, p. 26.

Lemmons, P. (1983). A Guided Tour of Visi On. *Byte, 1983*(6), pp. 256-278.

Lennox, J. (1985, August). Business With Pleasure. *What Micro*, pp. 15-18.

Leonard, A. (2000, May 16). *BSD Unix: Power to the people, from the code.* Retrieved May 24, 2013, from SALON: http://www.salon.com/2000/05/16/chapter_2_part_one/

Levy, S. (1999, August 29). *Behind The Gates Myth.* Retrieved from Newsweek: http://www.newsweek.com/behind-gates-myth-166002

Levy, S. (2004). *Insanely Great.* New York: Penguin Group.

Levy, S. (2010). *Hackers: heroes of the computer revolution.* Sebastopol, CA: O'Reilly.

Lewis, D. S. (1994, December 21). *A Discussion with Xerox PARC's Mark Weiser on the Future of Computing.* Retrieved August 16, 2013, from HOTT: http://virtualschool.edu/mon/SoftwareEngineering/MarkWeiserBckgroundForFuture

Libes, S. (1979, May). Radio Shack Has Over 50 Percent of Personal Computer Business. *Byte, 4*(5), p. 117.

Libes, S. (1995). *The Gary Kildall Legacy*. Retrieved November 19, 2011, from Digital Research: http://www.digitalresearch.biz/DR/Gary/newsx011.html

Lichstein, H. (1963, November 20). Telephone hackers active. *The Tech, 83*(24).

Linzmayer, O. W. (2004). *Apple Confidential 2.0: The Definitive History of the World's Most Colorful Company*. San Francisco, CA: No Starch Press.

Littman, J. (1993, September 12). The Last Hacker: He Called Himself Dark Dante. His Compulsion Led Him to Secret Files and, Eventually, the Bar of Justice. *Los Angeles Times*.

Liu, R. (1998, January 5). *Netscape has growing pains*. Retrieved from CNN Money: http://money.cnn.com/1998/01/05/technology/netscape/

Lohr, S. (2010, April 2). *H. Edward Roberts, PC Pioneer, Dies at 68*. Retrieved December 3, 2010, from The New York Times: http://www.nytimes.com/2010/04/03/business/03roberts.html

Los Angeles Times. (1995, April 18). Mac Clones Due in May. *Los Angeles Times*.

Mace, S. (1984, July 9). Amiga Demos its New Machine. *InfoWorld*, pp. 41-42.

Mace, S. (1985, June 3). Laser Printers Open Markets. *InfoWorld*, p. 44.

Mace, S., & Sorensen, K. (1986, May 5). Amiga, Atari Ready PC Emulators. *InfoWorld*, p. 5.

Maraia, V. (2005). *The Build Master: Microsoft's Software Configuration Management Best Practices*. Addison-Wesley.

Markoff, J. (1983, November 21). Microsoft Does Windows. *InfoWorld*.

Markoff, J. (1994, December 29). For Apple, Clones and Competition. *The New York Times*.

Markoff, J. (1996, December 23). Why Apple Sees Next as a Match Made in Heaven. *The New York Times*.

Markoff, J. (2005). *What the Dormouse Said*. Viking.

Martin, M. H. (1996, February 19). Paradise Lost. *Fortune*.

Matthews, I. (2003, February 15). *The Rise of MOS Technology & The 6502*. Retrieved December 7, 2010, from Commodore Computers: http://www.commodore.ca/history/company/mos/mos_technology.htm

McCarroll, T. (1984, December 24). Computers: A Flop Becomes a Hit. *Time*.

McIlroy, M. D., Morris, R., & Vyssotsky, V. A. (1971, June 29). *Darwin, a Game of Survival of the Fittest among Programs*. Retrieved April 2, 2010, from Dartmouth Computer Science: http://www.cs.dartmouth.edu/~doug/darwin.pdf

McTiernan, C. E. (1998). The ENIAC Patent. *IEEE Annals of the History of Computing, 20*(2), pp. 54-58.

Metcalfe, B. (1995, August 21). Microsoft and Netscape open some new fronts in escalating Web Wars. *InfoWorld*, p. 35.

Metcalfe, B. (1995, September 18). Without case of vapors, Netscape's tools will give Blackbird reason to squawk. *InfoWorld*, p. 111.

Microsoft. (1991, August). *Microsoft Windows Environment Version 3.1.* Retrieved from Tech Insider: http://tech-insider.org/windows/research/1991/1010.html

Microsoft. (1995, March 28). *Microsoft Announces Tools to Enable a New Generation of Interactive Multimedia Applications for The Microsoft Network.* Retrieved October 4, 2012, from Microsoft: ftp://ftp.microsoft.com/developr/drg/multimedia/Blackbird/BBPR.htm

Microsoft. (2001, May 3). *Accessibility with Responsibility.* Retrieved July 31, 2013, from Microsoft News Center: http://www.microsoft.com/en-us/news/features/2001/may01/05-03csm.aspx

Microsoft. (2006, November 2). *Microsoft and Novell Announce Broad Collaboration on Windows and Linux Interoperability and Support.* Retrieved July 13, 2013, from Microsoft News Center: http://www.microsoft.com/en-us/news/press/2006/nov06/11-02MSNovellPR.aspx

Microsoft. (2006). *Microsoft Corporation Annual Report 2006, Financial Highlights.* Retrieved December 21, 2012, from Microsoft: http://www.microsoft.com/investor/reports/ar06/staticversion/10k_fh_fin.html

Microsoft Corp. (1998, July 6). *Microsoft Posts Record Revenue and Income for Fiscal 1998.* Retrieved November 21, 2012, from Microsoft: https://www.microsoft.com/en-us/news/features/1998/7-16earnings.aspx

Microsoft Corporation. (2005, November 9). *Microsoft Corporation Annual Report 2005.* Retrieved May 24, 2013, from Microsoft: http://www.microsoft.com/investor/reports/ar05/downloads/MS_2005_AR.doc

Microsoft Limited. (1989). *IBM AND MICROSOFT EXPAND PARTNERSHIP; SET FUTURE DOS AND OS/2 DIRECTIONS.* London: Text 100.

Microsoft v. USA, 97-5343 (Court of Appeals for the District of Columbia April 21, 1998).

Miller, M. J. (1990, May 28). Windows 3.0: Worth the Wait. *InfoWorld, 12*(22), p. S1.

Miner, J. (1992, September). Jay Miner Interview. (M. Nelson, Interviewer)

Moeller, M., & Dodge, J. (1996, June 17). Netscape's Andreessen eyes 'Internet OS'. *PC Week.*

Money Programme, BBC2. (2011, December 14). Steve Jobs: Billion Dollar Hippy.

Moore, D. G. (1965, April 19). Cramming more components onto integrated circuits. *Electronics, 38*(8).

Moore, D. G. (1975). Progress in Digital Integrated Electronics. *Technical Digest*, pp. 11-13.

Moore, G. E. (1995, March 3). Interview with Gordon E. Moore. (R. Walker, Interviewer) Stanford University.

Moore, J. (Director). (2001). *Revolution OS* [Motion Picture].

Morris, P. R. (1990). *A history of the world semiconductor industry.* London: Peter Peregrinus Ltd.

Mosaic Communications. (1994, October 13). *Netscape, Available Now, Builds On Tradition of Freeware for the Net.* Retrieved October 2, 2012, from Mosaic Communications Corporation: http://home.mcom.com/info/newsrelease.html

Mott, T. (2010, June 4). *The Genius of Design 5: Objects of Desire.* BBC2.

Mueller, S., & Soper, M. E. (2001, June 8). *Microprocessor Types and Specifications.* Retrieved August 7, 2012, from informIT: http://www.informit.com/articles/article.aspx?p=130978&seqNum=4

Müller-Prove, M. (2002). *Vision and Reality of Hypertext and Graphical User Interfaces.* Universität Hamburg, Fachbereich Informatik.

Murdock, I. A. (1994, Jan 6). *Appendix A - The Debian Manifesto.* Retrieved July 16, 2013, from Debian: http://www.debian.org/doc/manuals/project-history/ap-manifesto.en.html

Narayen, S. (2011, October 6). *With Our Deepest Sympathy.* Retrieved 25 June, 2012, from Adobe Featured Blogs: http://blogs.adobe.com/conversations/2011/10/with-our-deepest-sympathy.html

Nelson, T. (1987). *Computer Lib.* Redmond, Washington, USA: Microsoft Press.

Nelson, T. (1987). *Dream Machines.* Redmond, Washington, USA: Microsoft Press.

Netcraft. (2013, July 2). *July 2013 Web Server Survey.* Retrieved July 18, 2013, from Netcraft: http://news.netcraft.com/archives/category/web-server-survey/

Netscape Communications. (1994, November 14). *MOSAIC COMMUNICATIONS CHANGES NAME TO "NETSCAPE COMMUNICATIONS CORPORATION".* Retrieved September 27, 2012, from holgermetzger.de: http://www.holgermetzger.de/netscape/NetscapeCommunicationsNewsRelease.htm

New York Times. (1985, October 1). Digital Research to Modify GEM.

New York Times. (1987, March 25). Company News; Atari, Commodore Settle. *New York Times.*

Newbart, D. (2001, June 1). Microsoft CEO take launch break with the Sun-Times. *Chicago Sun-Times*, p. 57.

Nicholson, M. (1986, October). Amstrad PC1512. *PC PLUS*(1).

Nocera, J. (1999, March 1). Witnesses in Wonderland On trial in Washington, Microsoft saw its witnesses get skewered, its video crash, and its prospects for victory take a serious turn for the worse. *Fortune Magazine*.

Noyce, R. N. (1959, July 30). *Patent No. 2,981,877*. United States of America.

Olmos, D. (1988, March 3). Chip Shortage Strains Computer Makers. *Los Angeles Times*.

O'Reilly, R. (1987, November 26). IBM Finds Key to Many New Rooms With OS/2. *Los Angeles Times*.

Parker, R. (1990, December 24). Two Giants With Own Views. *InfoWorld*, p. 8.

Parker, R. (1991, April 1). IBM's Marketing of OS/2 May Have Stunted Its Growth. *InfoWorld*, p. 38.

Parloff, R. (2007, May 14). *Microsoft takes on the free world*. Retrieved July 29, 2013, from CNNMoney: http://money.cnn.com/magazines/fortune/fortune_archive/2007/05/28/100033867/

Paterson, T. (2007, August 8). *Is DOS a Rip-Off of CP/M?* Retrieved November 19, 2011, from DosMan Drivel: http://dosmandrivel.blogspot.com/2007/08/is-dos-rip-off-of-cpm.html

PC User staff. (1989, January 14). Year in, Year out. *PC User*, p. 13.

PC's Limited. (1987). *Personal Computer Catalog*. Dell Computer Corporation.

Penenberg, A. L. (2009). *Viral Loop: The Power of Pass-it-on*. London: Hodder & Stoughton.

People's Computer Company. (n.d.). Retrieved September 6, 2010, from DigiBarn Computer Museum: http://www.digibarn.com/collections/newsletters/peoples-computer/

Petrosky, M. (1987, February 2). AppleShare airs at last. *Network World*, p. 4.

Pollack, A. (1990, March 24). Most of Xerox's Suit Against Apple Barred. *The New York Times*.

Poole, K. T. (2002). *Entrepreneurs and American Economic Growth: William H. Gates*. Retrieved June 2, 2012, from Voteview.com: http://voteview.com/gates.htm

Poor, V. (2004, December 8). Oral History of Victor (Vic) Poor. (G. Hendrie, & L. Shustek, Interviewers) Melbourne, Florida, USA: Computer History Museum.

Postel, J. (1981). *Request for Comments: 801*. Network Working Group.

Power, D. J. (2004, August 8). *A Brief History of Spreadsheets (v3.6)*. Retrieved November 29, 2011, from DSSResources.COM: http://dssresources.com/history/sshistory.html

Q-Success. (2013, July 18). *Usage of operating systems for websites.* Retrieved July 18, 2013, from W3Techs: Web Technology Surveys: http://w3techs.com/technologies/overview/operating_system/all

Quinlan, T., & Borzo, J. (1994, March 14). Power Macintosh banks on winning Windows users. *InfoWorld*, p. 1.

Ranney, K. (1986, January 20). Apple Planning 'Open' Mac. *InfoWorld, 8*(3), p. 1.

Raskin, J. (1979). *Computers by the Millions.* Retrieved April 4, 2012, from DigiBarn computer museum: http://www.digibarn.com/friends/jef-raskin/writings/millions.html

Raskin, J. (1979, May 29). *Design Considerations for an Anthropophilic Computer.* Retrieved April 5, 2012, from Making the Macintosh: http://library.stanford.edu/mac/primary/docs/bom/anthrophilic.html

Raskin, J. (1979, October 2). *Reply to Jobs, and Personal Motivation.* Retrieved April 5, 2012, from Making the Macintosh: http://library.stanford.edu/mac/primary/docs/bom/motive.html

Raymond, E. S. (Ed.). (1991). *The New Hacker's Dictionary.* London: The MIT Press.

Raymond, E. S. (1998, November 1). *The Halloween Documents.* Retrieved July 19, 2013, from Eric S. Raymond's Home Page: http://www.catb.org/~esr/halloween/

Raymond, E. S. (1999). *The Cathedral and the Bazaar.* Sebastopol, CA: O'Reilly & Associates.

Raymond, E. S. (2003). *The Art of Unix Programming.* Pearson Education.

Rebello, K. (1996, July 15). Inside Microsoft. *BusinessWeek.*

Rebello, K., & Arnst, C. (1992, November 29). 'The Great Digital Hope' Could Be A Heartbreaker. *Businessweek.*

Regis McKenna Public Relations. (1984, January 24). *Apple Introduces Macintosh Advanced Personal Computer.* Retrieved May 21, 2012, from http://library.stanford.edu/mac/primary/docs/pr1.html

Reid, B. (1993, September 28). *Alt Hierarchy History.* Retrieved September 12, 2012, from The Internet: http://www.livinginternet.com/u/ui_alt.htm

Reimer, J. (2005, December 14). *Total share: 30 years of personal computer market share figures.* Retrieved December 12, 2011, from ars technica: http://arstechnica.com/old/content/2005/12/total-share.ars

RESISTORS. (2009, December 16). Retrieved March 6, 2010, from http://resistors.org

Rheingold, H. (1985). *Tools for Thought.* New York: Simon & Schuster.

Ritchie, D. M. (1979, September). *The Evolution of the Unix Time-sharing System.* Retrieved May 24, 2013, from Alcatel-Lucent: http://cm.bell-labs.com/cm/cs/who/dmr/hist.html

Roberts, E. (1976, November). Charlatans, Rip-off Artists and Other Crooks. *Computer Notes*.

Rolander, T. (2007, August 9). The rest of the story behind Microsoft's OS deal with IBM. (R. Scoble, Interviewer)

Rolander, T. (2007, August 8). The rest of the story: How Bill Gates beat Gary Kildall in the OS war, Part 1. (R. Scoble, Interviewer)

Rose, F. (1989). *West of Eden: The End of Innocence at Apple Computer*. London: Century Hutchinson.

Rose, G. (2011, October 6). *Steve Jobs - here's to the crazy ones*. Retrieved from Virgin: http://www.virgin.com/news/steve-jobs-here%E2%80%99s-crazy-ones

Rosen, B. M. (1979, July 11). VisiCalc: Breaking the Personal Computer Software Bottleneck. *Morgan Stanley Electronics Letter*.

Rosoff, M. (2005, November). The Antitrust Price Tag. *Directions on Microsoft*, p. 46.

Roszak, T. (2000). *From Satori to Silicon Valley*. Retrieved August 19, 2010, from Making the Macintosh: http://library.stanford.edu/mac/primary/docs/satori/machines.html

Salkeld, R. (1981, June 16). *BBC Computer Literacy Project*. Retrieved from Retro Isle: http://www.retroisle.com/general/bbc_literacy_project.php

Salkever, A. (2004, September 30). *John, Paul, George, Ringo...and Steve?* Retrieved January 18, 2012, from Bloomberg Businessweek: http://www.businessweek.com/technology/content/sep2004/tc20040930_9317_tc056.htm

Salomon Brothers Inc. (1994). *Personal Computer Hardware*.

Salus, P. H. (1994, October). Unix at 25:The history of Unix is as much about collaboration as it is about technology. *BYTE*.

Satchell, S. (1986, December 22). ALR Access 386: Early Entry Performs Well Despite Shortcomings. *InfoWorld*, p. 49.

Sather, J. F. (1983). *Understanding the Apple II*. Chatsworth, CA: Quality Software.

Saxby, S. (1990). *The Age of Information: The Past Development and Future Significance of Computing and Communication*. Macmillan.

Scannell, E. (1988, September 12). Extended AT Bus Poses Threat to IBM. *InfoWorld*, p. 1.

Schaller, R. R. (2004). *Technological Innovation in the Semiconductor Industry: A Case Study of the International Technology Roadmap for Semiconductors (ITRS)*.

Schlender, B., & Woods, W. (1997, March 3). SOMETHING'S ROTTEN IN CUPERTINO AS CEO GIL AMELIO AND AN INEFFECTUAL BOARD DITHERED, APPLE COMPUTER LOST MARKET SHARE AND FADED INTO INSIGNIFICANCE. NOW

STEVE JOBS HAS RETURNED, WITH A TURNAROUND STRATEGY THAT COULD MAKE APPLE HIS ONCE AGAIN. *Fortune.*

Schmalensee, R. (1982, June). Another Look at Market Power. *Harvard Law Review.*

Schrage, M. (1984, January). Alan Kay's Magical Mystery Tour. *TWA Ambassador*, p. 36.

Schulman, A. (1993, September 1). Examining the Windows AARD Detection Code. *Dr. Dobb's.*

Schwartz, J. (2001, June 29). The Judge: A Court Overturned by an Appearance of Bias. *The New York Times.*

Scott, D. (1981, August). *Your Computer, 1*(2), p. 20.

Scott, J. (Director). (2005). *BBS The.Documentary* [Motion Picture].

Sculley, J. (2010, October 14). John Sculley on Steve Jobs. (L. Kahney, Interviewer)

Sculley, J. (2011, December 14). Steve Jobs: Billion Dollar Hippy. *Money Programme.* BBC2.

Segal, B. (1995, April). *A Short History of Internet Protocols at CERN*. Retrieved September 19, 2012, from http://ben.web.cern.ch/ben/TCPHIST.html

Seidner, R. (1996, June). Triumph of the Nerds: Part II. (R. X. Cringely, Interviewer) PBS.

Selmeier, B. (2006, February 6). *Digital Research Develops GEM for IBM PCjr.* Retrieved July 18, 2012, from Computer History Museum: http://corphist.computerhistory.org/corphist/view.php?s=events&id=1824

Shankland, S. (1999, August 11). *Red Hat shares triple in IPO.* Retrieved July 17, 2013, from CNET: http://news.cnet.com/2100-1001-229679.html

Simpson, R. (1997, September 15). A Talk with the Father of Computing. *Wired.*

Sinclair User. (1985, March). Sir Clive hits out in pub punch-up. *Sinclair User*(36).

Sjouwerman, S. (2006, November 20). Ballmer: "Linux Users Owe Me Money". *WServerNews, 11*(47).

Slater, R. (1987). *Portraits in Silicon.* Cambridge, Massachusetts: The MIT Press.

Slaton, J. (2001, December 13). *Remembering Community Memory.* Retrieved August 18, 2010, from SFGate.com: http://sfgate.com/cgi-bin/article.cgi?f=/g/a/2001/12/13/commmem.DTL

Solomon, L. (1984). *Solomon's Memory.* Retrieved July 2, 2010, from atariarchives.org: http://www.atariarchives.org/deli/solomons_memory.php

Spafford, E. H. (2009). *Quotable Spaf.* Retrieved September 13, 2012, from Gene Spafford's Personal Pages: http://spaf.cerias.purdue.edu/quotes.html

Speigelman, L. L. (1987, March 23). Developers Ready Mac SE-Compatible Portables. *InfoWorld*, p. 25.

Stallman, R. (1983, September 27). *new Unix implementation*. Retrieved June 27, 2013, from GNU Operating System: http://www.gnu.org/gnu/initial-announcement.html#f1

Stengel, S. (n.d.). *Old Computer Ads*. Retrieved November 24, 2011, from The Obsolete Technology Website: http://oldcomputers.net/oldads/old-computer-ads.html

Stephenson, N. (1999). *In the Beginning... Was the Command Line*. New York: HarperCollins.

Streibelt, P. (1992, June 27). *EASInet - IBM's Contribution to Scientific Networking in Europe*. Retrieved September 19, 2012, from http://www.caster.xhost.de/Sj92str.htm

Sun Microsystems. (1996, January 23). *JavaSoft Ships Java 1.0*. Retrieved from Tech Insider: http://tech-insider.org/java/research/1996/0123.html

Sun Microsystems. (2000, July 19). *SUN MICROSYSTEMS OPEN SOURCES STAROFFICE TECHNOLOGY*. Retrieved July 19, 2013, from Apache OpenOffice: http://www.openoffice.org/press/sun_release.html

Sketchpad (1962). [Motion Picture].

Sutter, H. (2005, March). The Free Lunch Is Over. *Dr. Dobb's Journal, 30*(3).

Swaine, M. (1997, April 1). Gary Kildall and Collegial Entrepreneurship. *Dr. Dobbs*.

Tesler, L. (1996, June). Triumph of the Nerds: Part III. (R. X. Cringely, Interviewer) PBS.

Tesler, L. (2010, June 4). *The Genius of Design 5: Objects of Desire*. BBC2.

The Modesto Bee. (1984, August 22). Atari sues over chips. *The Modesto Bee*, pp. D-11.

The National Science Board. (2000). *The Mansfield Amendment*. Retrieved November 25, 2013, from National Science Foundation: http://www.nsf.gov/nsb/documents/2000/nsb00215/nsb50/1970/mansfield.html

The New York Times. (1990, June 6). Windows 3.0 Sales At 100,000 Copies. *The New York Times*.

The New York Times. (1996, December 31). New Apple Chief got $3 million. *The New York Times*.

Thomas, J. (1997, September). Rhapsody in Redmond. *Developer Network Journal*, p. 14.

Thompson, T., & Baran, N. (1988). The NeXT Computer. *Byte*.

Thurrott, P. (2003, January 24). *Windows Server 2003: The Road To Gold, Part One: The Early Years*. Retrieved October 31, 2012, from Paul Thurrott's Supersite for Windows: http://www.winsupersite.com/print/windows-server/windows-server-2003-the-road-to-gold-part-one-the-early-years-127432

Tomczyk, M. S. (1984). *The Home Computer Wars: An Insider's Account of Commodore and Jack Tramiel.* Greensboro, North Carolina: COMPUTE! Publications.

Torvalds, L. B. (1991, August 26). *What would you like to see most in minix?* Retrieved June 28, 2013, from comp.os.minix.

Truss, L. (2003). *Eats, Shoots and Leaves: The Zero Tolerance Approach to Punctuation.* London: Profile Books.

Tynan, D. (2006, May 26). *The 25 Worst Tech Products of All Time.* Retrieved March 5, 2013, from PCWorld: http://www.pcworld.com/article/125772/worst_products_ever.html

University of California. (1986, March 4). *LICENSE AGREEMENT.* Retrieved 26 June, 2013, from Bell Labs: http://cm.bell-labs.com/cm/cs/who/dmr/bsdi/BSD_ATT_License.pdf

USA v. Microsoft, Civil Action 94-1564 (SS) (United States District Court for the District of Columbia July 15, 1994).

USA v. Microsoft, 98-1232 (United States District Court for the District of Columbia November 9, 1998).

USA v. Microsoft, 253 F.3d 34 (United States Court of Appeals for the District of Columbia Circuit June 28, 2001).

Valloppillil, V. (1998, August 11). *Open Source Software: A (New?) Development Methodology.*

van Hoesel, F. (1994, September 22). *Vatican Exhibit.* Retrieved September 24, 2012, from ibiblio: http://www.ibiblio.org/expo/vatican.exhibit/Vatican.exhibit.html

Veit, S. (2002). *How the Altair Began.* Retrieved July 2, 2010, from PC History: http://www.pc-history.org/altair.htm

Veit, S. (2002). *Processor Technology SOL.* Retrieved October 7, 2010, from PC History: http://www.pc-history.org/sol.htm

Visich, M., & Braun, L. (1974). *The Use of Computer Simulations in High School Curricula.* State University of New York at Stony Brook, Huntington Computer Project.

Wallace, J., & Erickson, J. (1992). *Hard Drive: Bill Gates and the Making of the Microsoft Empire.* New York: John Wiley & Sons.

Warnock, J. (1996, June). Triumph of the Nerds: Part III. (R. X. Cringely, Interviewer) PBS.

Warren, J. C. (1976, July). correspondence. *SIGPLAN Notices, 11*(7), 1-2.

Weber, J. (1993, June 19). Apple Computer's Sculley to Give Up CEO Position. *Los Angeles Times.*

Webster, R. (1987, December). Alan Kay: face values. *Personal Computer World,* pp. 128-135.

Weik, M. H. (1961, January). The ENIAC Story. *ORDNANCE.*

Weyhrich, S. (n.d.). *Most Popular Software Of 1978-1980.* Retrieved June 17, 2011, from Apple II History: http://apple2history.org/appendix/aha/aha78/

Weyhrich, S. (n.d.). *Storage Needs.* Retrieved October 6, 2011, from Apple II History: http://apple2history.org/history/ah05/

Williams, G. (1982, December). Lotus Development Corporation's 1-2-3. *BYTE*, pp. 182-198.

Williams, G. (1984, February). The Apple Macintosh Computer. *BYTE*, p. 30.

Williams, G., & Moore, R. (1984, December). The Apple Story Part 1, An Interview with Steve Wozniak. *BYTE*, p. A67.

Williams, G., & Moore, R. (1985, January). The Apple Story Part 2: An Interview with Steve Wozniak. *BYTE*, p. 167.

Williams, M., & Krazit, T. (2004, December 13). IBM Sells PC Unit to Lenovo. *InfoWorld*, p. 18.

Woerner, J. (2001, November 16). *The Story of the Datamath Calculator.* Retrieved May 19, 2010, from Datamath Calculator Museum: http://www.datamath.org/Story/Datamath.htm

Wolf, G. (1994, October). The (Second Phase of the) Revolution Has Begun. *Wired, 2*(10).

Wolf, G. (1995, June). The Curse of Xanadu. *Wired*(3.06).

Wozniak, S., & Smith, G. (2006). *iWoz.* London: Headline Review.

Yamagata, H. (1997, August 3). *Linus Torvalds Interview.* Retrieved July 1, 2013, from Hotwired Japan: http://web.archive.org/web/20061109173729/http:/hotwired.goo.ne.jp/matrix/9709/5_linus.html

Zachary, G. P., & Yamada, K. (1993, May 25). What's Next? Steve Jobs's Vision, So on Target at Apple, Now is Falling Short. *Wall Street Journal.*

Zakon, R. H. (2011, December 30). *Hobbes' Internet Timeline 10.2.* Retrieved September 18, 2012, from Robert H'obbes' Zakon: http://www.zakon.org/robert/internet/timeline/

Zaks, R. (1981). *From Chips to Systems: An Introduction to Microprocessors.* Sybex.

Index

Page numbers in **bold** indicate primary or explanatory coverage. Page numbers in *italics* indicate an entry in the Timeline section.

Lightning Source UK Ltd.
Milton Keynes UK
UKOW03f0610260614